DIGITAL ELECTRONICS AND LABORATORY COMPUTER EXPERIMENTS

DIGITAL ELECTRONICS AND LABORATORY COMPUTER EXPERIMENTS

Charles L. Wilkins
University of Nebraska

Sam P. Perone
Purdue University

Charles E. Klopfenstein
University of Oregon

Robert C. Williams
University of Nebraska

Donald E. Jones
Western Maryland College

PLENUM PRESS • NEW YORK AND LONDON

Library of Congress Cataloging in Publication Data

Main entry under title:

Digital electronics and laboratory computer experiments.

Bibliography: p.

Includes index.

1. Electronic data processing–Chemistry. 2. Electronic digital computers. I. Wilkins, Charles L.

QD39.3.E46D53 542'.8 75-11570

ISBN 0-306-30822-3

Plenum Press, New York is a division of Plenum Publishing Corporation
227 West 17th Street, New York, N.Y. 10011

United Kingdom edition published by Plenum Press, London
A Division of Plenum Publishing Company, Ltd.
Davis House (4th Floor), 8 Scrubs Lane, Harlesden, London, NW10 6SE, England

This material was prepared with the support of National Science Foundation Grants GJ-441, GJ-393, and GJ-428. Any opinions, findings, conclusions or recommendations expressed herein are those of the Authors and do not necessarily reflect the views of the National Science Foundation

Printed in the United States of America

Preface

Science undergraduates have come to accept the use of computers as commonplace. The daily use of portable sophisticated electronic calculators (some of them rivaling general-purpose minicomputers in their capabilities) has hastened this development. Over the past several years, computer-assisted experimentation has assumed an important role in the experimental laboratory. Mini- and microcomputer systems have become an important part of the physical scientist's array of analytical instruments. Prompted by our belief that this was an inevitable development, we began several years ago to develop the curricular materials presented in this manual.

At the outset, several objectives seemed important to us. First, insofar as possible, the experiments included should be thoroughly tested and error free. Second, they should be compatible with a variety of laboratory-computer, data-acquisition, and control systems. Third, little or no previous background in either electronics or programming should be necessary. (Of course, such background would be advantageous.) To satisfy these objectives, we decided to adopt a widespread high-level computer language, BASIC, suitably modified for the purpose. Furthermore, we have purposely avoided specifying any particular system or equipment. Rather, the *functional characteristics* of both hardware and software required are stipulated. The experiments have been developed using Varian 620 and Hewlett-Packard 2100 series computers, but we believe they are readily transferable to other commonly available computer systems with a minimum of difficulty.

The entire set of experiments can be used as a basis for a laboratory course. If this is done, we suggest that all the digital-electronics experiments (Experiments 1–5) be performed, as the topics covered in those experiments are essential for success in the subsequent ones. (Experiment 5 is the single exception to this suggestion; because it is quite lengthy, Secs. C and D may optionally be replaced by, or combined with, Experiment 7. However, Secs. A and B should definitely be assigned.) Experiments 6, 7, and 8 form a useful bridge between the digital electronics and the computer experimentation and should probably also be considered mandatory. Following that, a selection of several from the remaining experiments completes the logical progression from the simple building blocks of electronics and programming to the real-world kinds of problems where the computer can play an important role. Most of the experiments can be completed in one four- or five-hour laboratory period, or two three-hour periods. If the latter arrangement is used, the experiments contain convenient stopping spots.

Obviously, depending on the level of the students' knowledge, selected experiments may be incorporated in existing physical or analytical chemistry courses if time and interest do not permit a full course dealing with laboratory-computer applications. In any case, we believe the experiments in this manual can provide students with a rewarding and challenging introduction to these techniques.

Our students over the years have played an essential role in the development and testing of the experiments included here. We would like to acknowledge by name the most important contributors and to thank collectively those many individuals who made lesser, but equally worthwhile, contributions. Accordingly, we thank James Evans, James Koskinen, A. Randall Heap, and R. Bradford Spencer (University of Oregon); John Eagleston, David O. Jones, Frank Pater, and Keith Dahnke (Purdue University); and Joseph Norbeck, Ray Swanson, David Thoennes, Douglas Weiss, Brian Dodson, and Ted Stout (University of Nebraska). Finally, we are particularly appreciative of the typing efforts of Mrs. Dea Gustafson.

The financial support of the National Science Foundation through grants GJ-441, GJ-393, and GJ-428 is also gratefully acknowledged.

Charles L. Wilkins
University of Nebraska

Sam P. Perone
Purdue University

Charles E. Klopfenstein
University of Oregon

Robert C. Williams*
University of Nebraska

Donald E. Jones
Western Maryland College

*Present address: Central Research Laboratories, 3M Company, St. Paul, Minnesota.

Contents

Section III. Appendices

Introduction

HISTORICAL DEVELOPMENT OF LABORATORY COMPUTERS

The purpose of this laboratory manual is to introduce you to the techniques which are available for making use of the considerable capabilities of the laboratory computer in experimental science. Before we launch into a detailed discussion of experiments and methods required to make use of that technology, a brief consideration of some of the fundamental elements of the general-purpose computer and the historical development of its use in the laboratory is in order.

Because the earliest electronic digital computers became available in 1953, the history of their use is brief. Even so, that use has seen enormous change and progress in the relatively brief span of years in which it has been possible to use computers. Most scientists are well aware of the enormous speed with which digital computers can carry out complicated calculations. Because of their extensive use as computational machines, computers have become most familiar to the average person as a means for keeping accounts straight (his acquaintance coming in the form of computer-generated bills) or carrying out complicated mathematical calculations in a small fraction of the time which would otherwise be required. This image of the computer as essentially a computational device has, for many, obscured another related and very important property of the digital computer. This is its capability to make very rapid control decisions or to communicate at high speeds with other types of devices, such as laboratory instruments. It is, of course, obvious that these are simply consequences of the high computational speed possible. In this laboratory manual, we will focus our attention primarily upon those capabilities which make possible the performance of laboratory experiments more rapidly or more precisely than would otherwise be feasible, or the performance of entirely new types of experiments that could not be carried out without the help of the computer.

Historically, scientists first became aware of the potentialities of computer use in the laboratory in the early 1960's. At that time, computers were still relatively expensive and probably not economically feasible for any but the largest, most well-funded laboratories. However, scientists became aware that, if data could be transmitted to a computer in digital form, the maximum amount of information could be extracted from experiments in the minimum length of time. Early applications of digital-computer methods in the chemistry laboratory focused on data-logging techniques, whereby some type of digital or analog data-acquisition device was attached to an experiment and the data collected later transferred to the computer for reduction and analysis. Sometimes the data were collected directly into a device such as a multichannel analyzer and then transferred to punched cards or paper tape for computer analysis. It was during this period that the computer-averaged transient (CAT) technique was introduced for nuclear magnetic resonance spectrometry. In this application, the ratio of signal to noise for an NMR spectrum was enhanced by the performance of multiple experiments with the data added coherently into an analyzer. This permitted subsequent display of the "time-averaged" spectrum on an oscilloscope or plotter at the termination of the experiment. This method gave workers in NMR spectrometry a hint of the great power which the digital computer could bring to their field of interest. Direct recording of analog information on magnetic tape, with subsequent conversion to digital form which a general-purpose digital computer could deal with, was also widely used at this time. Workers in the field of mass spectrometry developed methods of recording mass spectra on photographic film and then automatically scanning the film with a suitable microdensitometer to measure peak positions and intensities. These data, recorded on magnetic tape, could then be further analyzed by a digital computer.

Although data-logging methods did allow experimenters to make more efficient use of digital computers and to gain some advantage in treatment of the experimental data, it soon became evident that the weak link in the procedure was the transmission of the data to the general-purpose computer. It became obvious that a more desirable situation would be direct entry of data from the experiment into the computer, if possible. Unfortunately, in the early days of such use, the expense of digital computers was very high, and it was rare that the importance of an experiment could justify dedication of a computer to its needs. Furthermore, construction of high-speed data-transmission lines and the necessary electronic equipment to allow transmission of data from the experiment site to the computer was also fairly expensive. Consequently, the introduction of the computer to the laboratory, in the broadest sense, did not begin until advances in electronics technology brought the cost of computers down to a small fraction of their earlier cost. At this point, it became practical to consider a great many other ways to use the computer in the laboratory. We will now turn to a consideration of precisely how that has taken place.

LABORATORY APPLICATIONS OF DIGITAL COMPUTERS

As mentioned above, the earliest use of computers for treatment of laboratory data was simply to analyze and reduce the data which had been previously collected by data-logging devices. This, in itself, was a marked aid to the experimenter and analyst, but still fell short of the ultimate desired objective—fully automatic data collection and reduction. When prices of laboratory computers had fallen to levels consistent with their use in the laboratory, the first step taken by many laboratory scientists was to obtain small computers (now commonly called minicomputers) equipped for direct data acquisition from laboratory instruments. These, along with suitable devices for listing or display of the data thus obtained, comprised relatively simple, yet powerful, laboratory computer systems. In this way, a minimal digital computer could collect information from an experiment directly and either preserve it on a storage medium, as was done previously with the logging devices, or display it after transforming it in suitable ways. Clearly, this sort of logging application was somewhat superior to the previous methods in that computations using the data collected could be conveniently performed *on-line* (that is, while the computer was attached to the experiment), but still the full range of possibilities was not implemented. The next step in the logical progression of laboratory computer use was design and construction of *interfaces* (devices allowing two-way communication between computer and experiment) to permit control of the experiment by the digital computer. Now, in addition to its data acquisition and reduction functions, the computer was able to carry out a preprogrammed sequence of operations to achieve the desired experimental results. In its simplest form, this sort of control is called *open-loop control.* This latter term refers to control situations where a preselected sequence of control instructions is carried out, and there is essentially no interactive feedback involved. A somewhat more sophisticated type of control places the experiment, the computer, and the experimenter in a loop where each can respond to various experimental conditions and modify the control sequence in the process. This sort of control, called *closed-loop control,* provides a means for constant optimization of experimental conditions if the design of the operating system is carefully handled.

One additional, highly desirable use of a laboratory computer is to handle files of data. This sort of file manipulation potentially can be used in many ways. For one thing, reference files of spectral or other data may be kept on hand in computer-readable form and compared with similar data obtained from unknown materials for analysis or identification purposes. Alternately, files of data may simply be stored for future reference much in the way definitions of words are stored in a dictionary. In this sort of application, information of interest to the scientist in the laboratory may be made readily available and, at the same time, compactly stored. The ultimate objective, often not realized, is a computer system which combines all of these fundamental capabilities and permits the experimenter to conduct an entire experiment from beginning to end under complete computer control (with the possibility of the experimenter intervening during the process), ultimately to yield an experiment report which concisely summarizes all important results and information obtained from the experiment. In practice, the various aspects of this objective are realized in differing degrees; but, in any case, the end result of the introduction of the digital computer to the scientific laboratory is generally an enhancement of the experimenter's capabilities and an improvement in both quality and quantity of data which can be obtained and processed.

LABORATORY COMPUTER CHARACTERISTICS

One of the significant factors which made it possible to introduce a general-purpose digital computer to the laboratory was the technological state of the art. In other words, as computer-manufacturing techniques became more and more efficient, the cost of component parts and, hence, the computer itself decreased. Now, instead of costing many times what the experimental apparatus itself might cost, a reasonably sophisticated laboratory computer can often be purchased for a fraction of the cost of the experimental apparatus with which it is to work. When this is the case, it makes sense, if the computer can increase the experiment throughput or quality of the data, to invest in a laboratory computer as an *accessory* to the experimental instrument in question.

What are the properties which make this possible? Are the capabilities required for a laboratory computer system any different from those required for a purely computational application? Basically, the computer itself need not be much different from one required for computational use only. The main difference between laboratory computer systems and others is the addition of the requisite elements for the communication between the computer and laboratory instruments. In a very real sense, the problem involved here is no different from the problem involved whenever any input/output device is attached to a computer. From a computer scientist's standpoint the laboratory instruments may very well be thought of as special-purpose input/output devices for the computer. From the laboratory scientist's point of view, the computer is simply an accessory which allows him to make better use of his instruments. The elements required to make this possible are devices which permit analog information (i.e., voltages, currents, or mechanical motion) to be translated into digital form which the computer can store

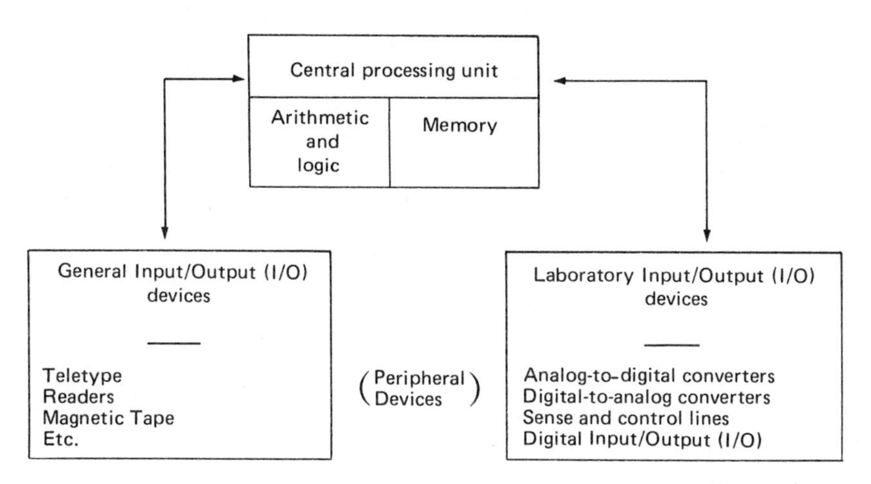

Fig. 0-1
Block diagram of a general-purpose laboratory computer system.

and manipulate, digital-to-analog converters which allow the reverse kind of translation so that the information stored in the computer can be communicated in analog form to the experiment or instrument in question, and digital sense, control, or input/output lines. These constitute the fundamental array of hardware needs for reasonably complete computer–instrument communication. We will consider all of these elements in greater detail later in the manual.

Functionally, a general-purpose laboratory computer system can be described by a simple block diagram such as that in Fig. 0-1. The basic elements of the computer itself are simple and easily understood. First, there is the section which contains the electronic hardware to carry out *arithmetic* and *logic* operations. Second, there is the computer *memory.* This serves the function of storing either data or programs, generally in a way which allows both to be modified at will. Finally, there is the *central processing unit* (or the "brain") of the computer which controls the movement of data and instructions from one to the other of these elements and which usually also controls movement of data to or from the outside world through the input and output devices. We have not mentioned input and output devices as a fundamental part of the computer because, important as they are, they are external to the computer itself and are part of a group which are collectively known as *peripheral devices.* The most common such device is the Teletype. This allows communication between the computer operator and the computer. There is a very large array of such devices which are devoted basically to allowing the input of digital information in the form of numeric or alphabetic characters and the output of similar information. These include high-speed line printers, card punches, cathode-ray-tube devices, and paper tape punches and readers. Magnetic disk and magnetic tapes are extensively used for both temporary and semipermanent storage of data and programs. Some of these so-called peripheral devices (for example, high-speed line printers and card punches) are so expensive as to make their use prohibitive in conjunction with inexpensive laboratory computers. Although prices for these latter items are decreasing, it is still true that either one of the two mentioned can easily cost more than the minicomputer itself. Fortunately, the requirements of the laboratory are such that card punches and high-speed line printers are not necessary, although they might very well be useful.

Special purpose input/output devices such as analog-to-digital converters, digital-to-analog converters, sense and control lines, and, in many cases, interrupt lines are really the most important parts of a laboratory computer system. These devices are not particularly expensive in relation to the cost of the computer, and they are certainly a necessity if data acquisition and control of the sort we mentioned above are to be effectively implemented. It is the purpose of this laboratory manual to introduce you, through a series of experiments, to the use of such devices and teach you the required techniques and principles for most efficient use of this hardware. As a result, completion of the experiments included in the manual should allow you to make use of the capabilities of the digital computer in connection with your own experiments in the laboratory. In the next section of this introduction, we will briefly consider the organization of the laboratory manual and the purpose of the various types of experiments included.

LABORATORY COMPUTER TRAINING APPROACH

In our experience in training chemists of widely varying computer backgrounds to take advantage of laboratory computers, we became aware that it is necessary to lead students sequentially from a simple lecture and laboratory introduction to fundamental principles of computer interfacing. At the same time, it is equally important to consider not only the electronic hardware, but the programming (or software) problems which must be met and solved in order to effectively use laboratory computers. Central to this latter aspect of the training is an understanding of some of the principles of information theory and the ways in which digital techniques can be employed to make advantageous use of the computer's capabilities. Our guiding philosophy is that the experiments should be kept at a level such that student background requirements are minimized (i.e., no previous programming or electronics background should be required), and that the experiments should teach the fundamentals of chemical-instrumental or data-evaluation principles involving computerization. In order to meet these objectives in a way consistent with our philosophy, we use the high-level algebraic language, BASIC,* throughout the manual and present the experiments in such a way that extrapolation to a similar language (e.g., FOCAL) will be simple. The reason for this choice is threefold. First, the language is *simple* and readily learned (since it was designed for student use). Second, its *interactive interpretive* nature allows relatively easy program development and correction. Third, and especially important, this language is *available* on the majority of laboratory computers in use, and, for a great many applications, the direct extrapolation of procedures developed in the experiments included in this manual will be suitable for research applications. We have tried to make the experiments as independent of a particular type of laboratory computer system as possible, and the use of BASIC has helped us achieve that end. All the experiments presented have been extensively tested with students in our own courses. We know that they are feasible for junior/senior level students, and we are reasonably confident that they are as error free as possible. We have actually developed the experiments using Hewlett-Packard and Varian computers, together with versions of BASIC which operate with those systems. The appendices at the end of the manual contain a detailed description of both types of systems upon which the experiments were developed.

It is our belief that this approach confers sufficient generality upon the manual that the experiments can be adapted to almost any contemporary laboratory computer system with a minimum of effort.

The experiments are grouped in two categories. The first of these categories, *Logic Experiments*, is intended to introduce the beginning student to the properties of simple digital logic elements, the principles involved in their use, the construction and use of more complex functions from simple logic elements, and the properties and use of digital-to-analog and analog-to-digital converters. Completion of this sequence of five experiments will provide the necessary background for the more advanced computer-interfacing experiments. For those students who are already familiar with the principles and applications of digital logic, this section and group of experiments provide a valuable review. For students who have not been previously introduced to digital logic, this group of experiments is essential to successful completion of the later experiments. We expect, of course, that students with desire for more advanced knowledge of this field will consult one of the several excellent texts available, a partial list of which is given in the Bibliography.

The second major category of experiments, *Computer Interfacing Experiments*, consists of a variety of experiments intended to expose the

*See Sec. B of the Bibliography.

student to a representative sampling of the types of problems which must be faced in attempting to utilize the computer in the scientific laboratory. Simple data-acquisition experiments are followed by ensemble averaging and multiplexing experiments which involve important concepts in timing and synchronization. A group of experiments on simulation and graphic display provide the student with the necessary background to carry out more sophisticated display experiments which involve actual computer-assisted chemistry applications covering a range of applications. Computer-aided potentiometric titrations, gas chromatography, kinetic analysis, and spectrophotometry are examples of the real applications contained in later experiments.

In conclusion, we believe that the diligent student who carefully studies the introductory material provided for each major group of experiments and performs and understands the experiments will have obtained a reasonable introduction to laboratory computing. Furthermore, he will be in a position to make practical use of this knowledge for experiments of his own devising. The requisite element for achieving these objectives is a suitable laboratory computer system with capabilities for data acquisition from, and control of, experimental devices.

A high-level language such as BASIC permits students who perform the experiments in the manual to achieve working programs with a minimum of study of computer programming. It is to be expected that, eventually, especially in the later experiments, some limitations in the BASIC language may become apparent. At that time, some students may find it worthwhile to investigate the technique of assembly language programming.

By means of flow charts and (occasionally) sample programs, the manual illustrates reasonable approaches to solving many of the problems posed by the experiments. It is not our intent to provide exhaustive, specific detailed instructions for all aspects of the experiments since to do so would rob the student of much of the fun of exercising his own initiative to achieve successful working interfaces to the computer. We believe that the approach outlined here will bring the student to a position where he not only understands and appreciates the use of laboratory computers but can actually make effective practical use of these devices on a routine basis. He will also be in a position to further develop more sophisticated approaches to laboratory computing problems than are possible in a manual of this sort.

Section I

Experiments 1–5: Principles of Digital Logic

Section I

Introduction to Experiments 1–5: Principles of Digital Logic

Experimental scientists usually consider the principles of digital logic to be too complicated to be easily understood without extensive study and training. However, those who do venture into the area find that these principles are based on simple rules, and the "hocus-pocus" often associated with computer operations is missing. They find that most of the principles are already applied in various forms throughout everyday life, and that binary operations in logic circuits are no more complicated than electrical wiring in a home. Light switches, after all, are binary control devices used to set the logical condition of a lamp. Similarly, a door bell switch is a trigger for another binary event, the striking of a gong. These comparisons can be carried even further, in that a lamp switch is usually "latching," that is, positive action is required to complement or change the logical state of the switch. The lamp stays on, or off, until the switch position is changed. Doorbell activators, on the other hand, provide only a pulse of information each time the switch is pushed. Here, already, we have found rather simple illustrations for two important concepts in digital logic, that of bistable (or latching) and monostable (or one-shot) circuits. Other equally simple comparisons will be used in the following experiments to explain the functions of various elements in digital logic design.

In the experiments, we have made nearly exclusive use of integrated circuit (IC) packages for implementing the required logical functions. It is really this striking advance in technology that has brought advanced digital capabilities into our teaching and research laboratories. The scientist implementing an experiment no longer needs to be concerned with whatever transistors, resistors, etc., are required to implement a logical function, since everything is now provided in small, inexpensive plastic packages. Designers of interface circuits need only be concerned with the simpler task of understanding the functional characteristics of the entire device. The experiments in this chapter are designed to provide the practice and experience needed for practical application of digital logic in the laboratory.

Digital-logic design and construction is simple to understand and easy to implement. No complicated formulas or concepts need be memorized, and certainly no extensive background in electronics is required for successful implementation of most logic circuits.

The basic element of today's logic is the *b*inary dig*it* (*bit*). This logical unit of information is either "true" or "false," with "true" usually represented as a 1 and "false" as a 0. Electrically the bit can be represented by a voltage level, or as a current level, depending on the component used. Conventions adopted for a particular type of logic (each type will be discussed later) are standard throughout the semiconductor industry, so that, after learning a handful of rules, it is possible to apply them to a broad range of functional designs. At last count, more than 200 different varieties of starting modules were available, each making the job of the designer easier.

It is perhaps easiest to grasp the notions of digital logic by considering the binary action of a common electrical switch.

In Fig. I-1a, no voltage appears at the output because the switch is in an open (or binary 0) state. Closing the switch, as in Fig. I-1b, allows electrical contact between the input and output, and, therefore, the voltage out is equal to the voltage in (logical 1). We can expand this concept by adding switches to the output side of the examples above, as in Figs. I-2 and I-3. In Fig. I-2, voltage appears at the output if *either* switch A OR B is closed. In Fig. I-3, we see that both switches A AND B must be closed for a voltage to appear at the output. Now, if we replace the switches with electronically controlled devices that pass current when stimulated with voltage, and are otherwise insulators, we have constructed the two basic elements of digital logic: the AND and OR gates. The actual voltages used will vary according to the type of electrical components used. In the commonly used transistor–transistor logic (TTL), transistors replace the switches above and logical true (1) is represented by

Fig. I-1
Electric switch functions: (a) binary zero; (b) binary one.

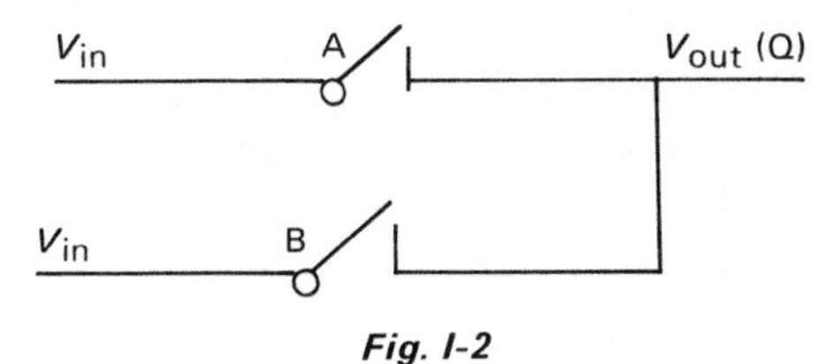

Fig. I-2
An electric switch OR circuit.

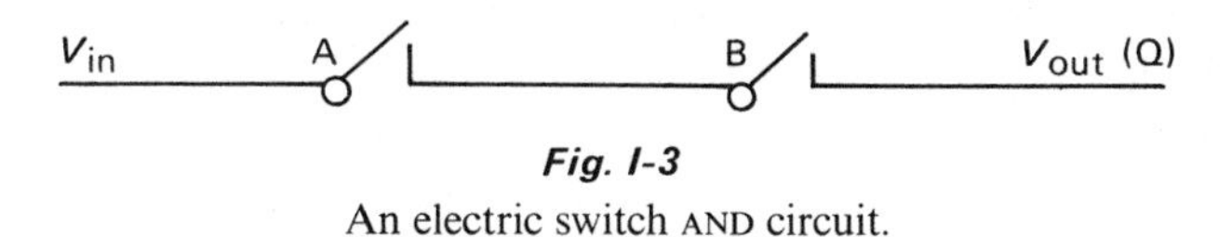

Fig. I-3
An electric switch AND circuit.

+3 to +5 V, while logical false (0) is 0 to 0.3 V. Later we will perform some experiments to measure these levels for particular devices (see Experiment 1).

Figure I-4 gives the usual symbols used to represent the AND and OR functions in logic diagrams, along with the "truth tables" for the devices, which will be discussed in greater detail in Experiment 1. With these functions, one might think that nearly any function could be implemented in hardware form. However, one basic element is missing. That is the capability to negate or "complement" the logical sense of a signal. Complementation (also called inversion) is easily implemented with transistors, and this function is represented by the symbol given in Fig. I-5.

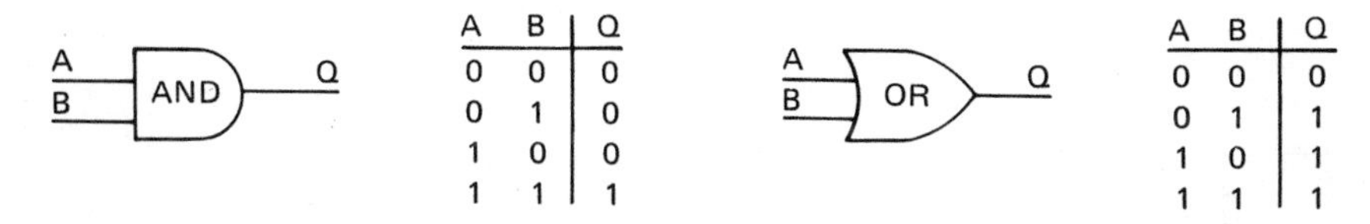

A	B	Q
0	0	0
0	1	0
1	0	0
1	1	1

A	B	Q
0	0	0
0	1	1
1	0	1
1	1	1

Fig. I-4
Symbols and truth tables for logical AND and OR functions.

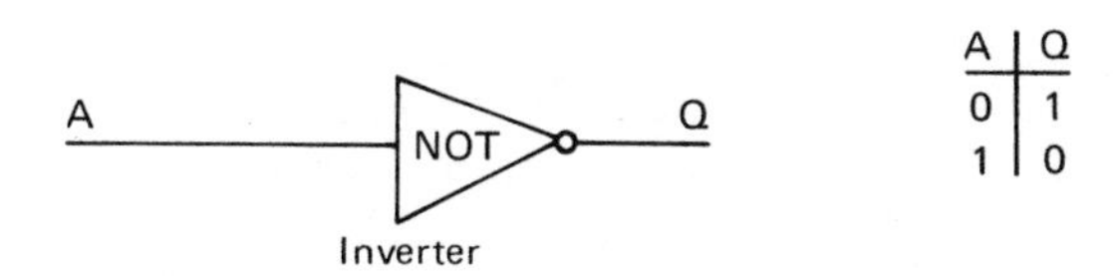

A	Q
0	1
1	0

Fig. I-5
Symbol and truth table for an inverter.

When an inverter and AND or OR gate are combined, the resulting complementary functions are termed NAND (NOT AND) and NOR (NOT OR). Simplified symbols are used to represent these functions as given in Fig. I-6. It is also possible to generate the NAND and NOR functions in a less direct manner using a logic identity known as DeMorgan's theorem. DeMorgan's theorem can be stated in two alternative forms: 1) NOT A OR NOT B ≡ NOT (A AND B); 2) NOT A AND NOT B ≡ NOT (A OR B). The logic-symbol equivalents of this expression are shown in Fig. I-7. (The reader is referred to the references for a more detailed discussion of DeMorgan's theorem.)

By combining these functions, any logical expression may be implemented. Using DeMorgan's theorem, it is even possible (but not necessarily desirable) to implement all logical functions using just NAND gates. (Note that an inverter can be considered to be a one-input NAND gate or, alternately, a one-input NOR gate.) Although it is not efficient to rely on any one type of gate, NAND and NOR gates are used far more commonly, principally because they have slightly better standards of performance. The AND–OR functions are readily generated when necessary by complementing NAND–NOR functions.

The final element we will discuss here is a bistable circuit that can be used for storing binary information. In Fig. I-8, A and B are held at logical 1 for as long as data storage is required. If A is momentarily lowered to logical 0, then restored to logical 1, output Q will be set to logical true. The same operation, performed on B, will cause Q to become false. We will see later cases where this circuit is used to store flag commands by the computer which are reset by experimental devices.

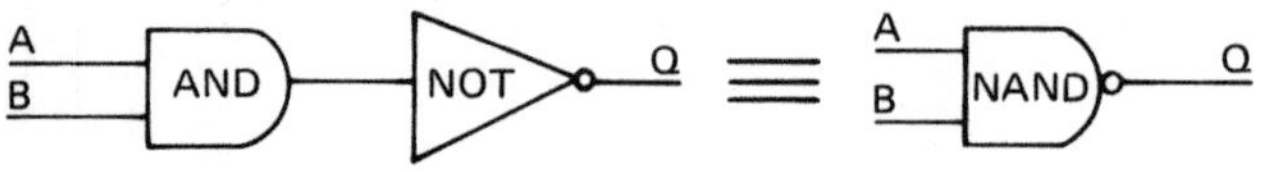

A	B	Q
0	0	1
0	1	1
1	0	1
1	1	0

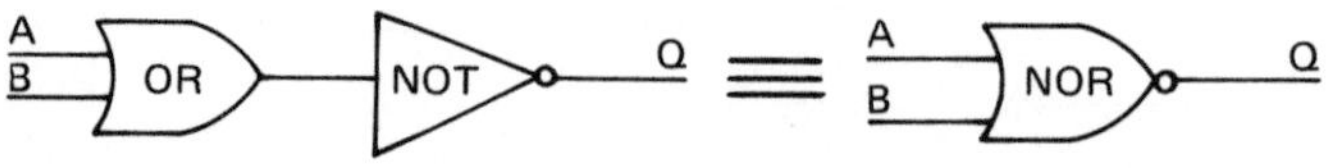

A	B	Q
0	0	1
0	1	0
1	0	0
1	1	0

Fig. I-6
Symbols and truth tables for logical functions NAND and NOR.

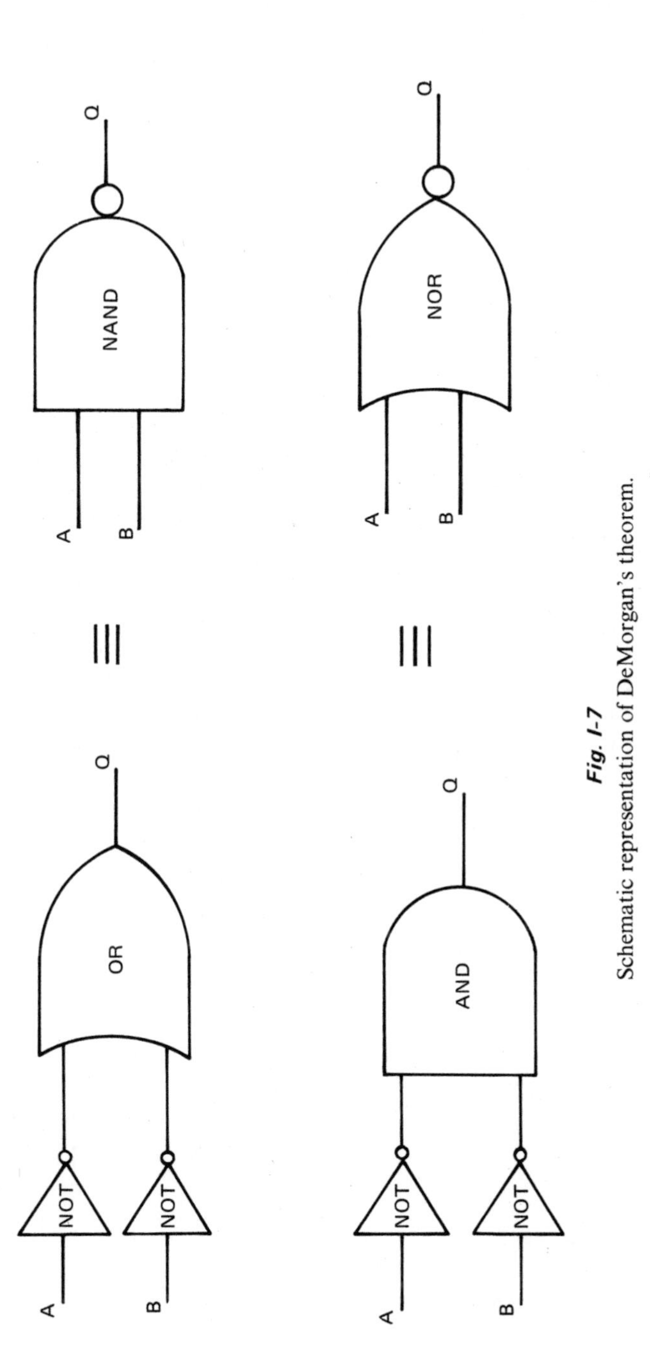

Fig. I-7
Schematic representation of DeMorgan's theorem.

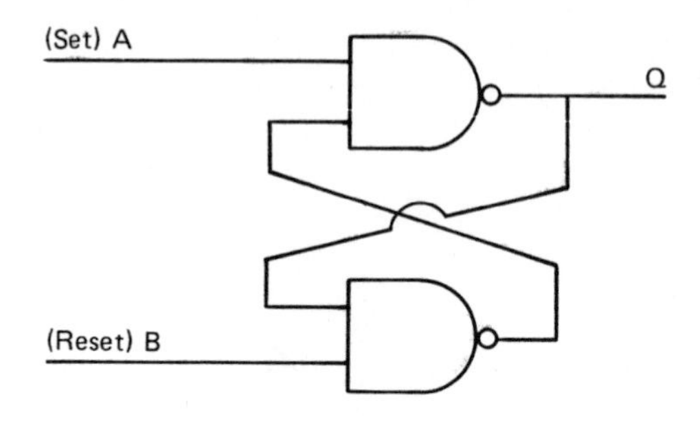

Fig. I-8
A circuit capable of storing binary information.

LOGIC TYPES

Digital-logic-circuit designs have evolved into four or five basic types. The variations include modes of input, output, and semiconductor techniques used to implement the logic. Some types are compatible with circuits from other classes, some are not.

The most common, by far, of all the various styles of logic is the family resistor–transistor, diode transistor, and transistor–transistor logic (RTL, DTL, and TTL). These differ primarily in the modes of input used and all share common output modes and power-supply requirements. The devices are increasingly complex, and hence were developed progressively in time to the point where, of these three, TTL is most commonly used today. The reader is referred to more advanced texts for details (see Bibliography); however, the schematics in Fig. I-9 illustrate the functional differences

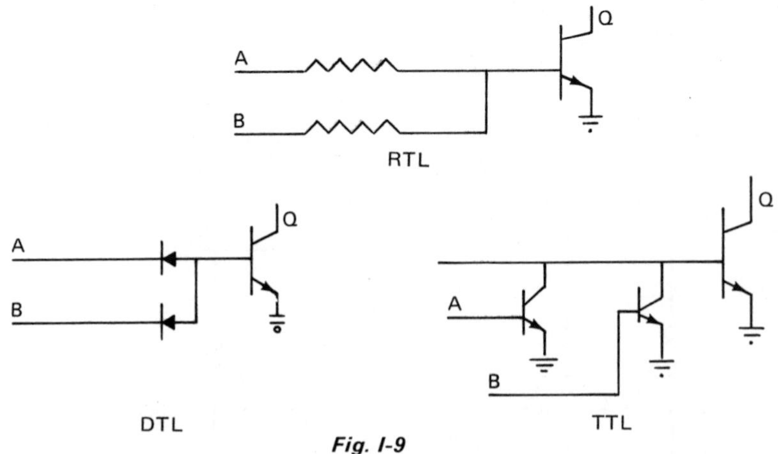

Fig. I-9
RTL, DTL, and TTL logic types.

among the classes, leaving out biasing circuitry and other factors unnecessary for this discussion. Logically, the function implemented in Fig. I-9 is identical to that provided by switches in Fig. I-2, except that due to the action of the transistor, the signal is complemented on output. The function is NOR, as described in Fig. I-6. RTL and DTL, although simpler to construct, do not provide the same isolation and drive characteristics as TTL.

DIGITAL LOGIC AND THE COMPUTER

The devices that allow communication between computers and experiments are called interfaces. These elements are made up of the components described above to form digital registers for input and output of binary information. Digital-to-analog converters and analog-to-digital converters perform the conversion functions implied and are interfaced to the computer via the binary paths described above. Today, these devices are available in inexpensive modular form from a number of suppliers. The diagram in Fig. I-10 illustrates how these different elements are combined to form a complete system. A more detailed description of the modular interface components is given in Appendix C.

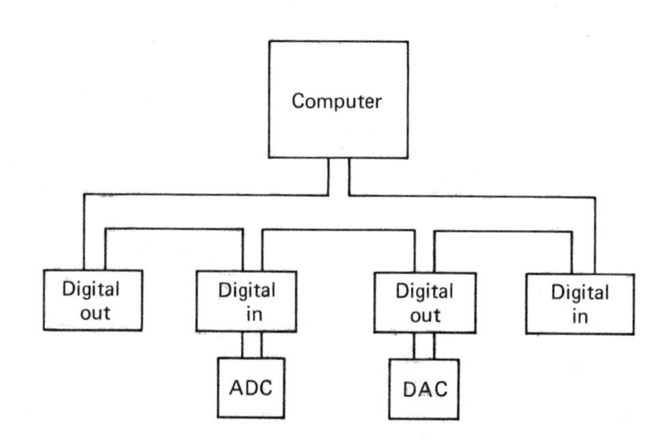

Fig. I-10
Block diagram for a computer, equipped with digital interfaces.

Experiment 1

DC Properties of Logic Elements

In order for a computer to be able to monitor an experiment, the instrumentation must be tied in with the computer. Usually, a continuous (analog) signal must be monitored, and some sequence of events must be controlled by the computer (turn on a switch, start a clock, store data, calculate values, etc.). This sequence of events can be caused to occur by an appropriate combination of logic elements. These logic elements may be rather simple or highly complex. The simplest of these is called a *gate* and may be considered as a simple electronic on–off switch. When the switch is closed, a signal goes through; when open, it does not. Several of these on–off switches can be combined in such a way that they simulate simple logical expressions.

Inspection of Figs. 1-1 and 1-2 shows two such configurations.

In Fig. 1-1, if a standard (nonzero) voltage is applied at point V, both switch A AND switch B must be closed for point Q to reach that voltage. When this occurs, point Q is considered at a logical "1" or "True" state. If

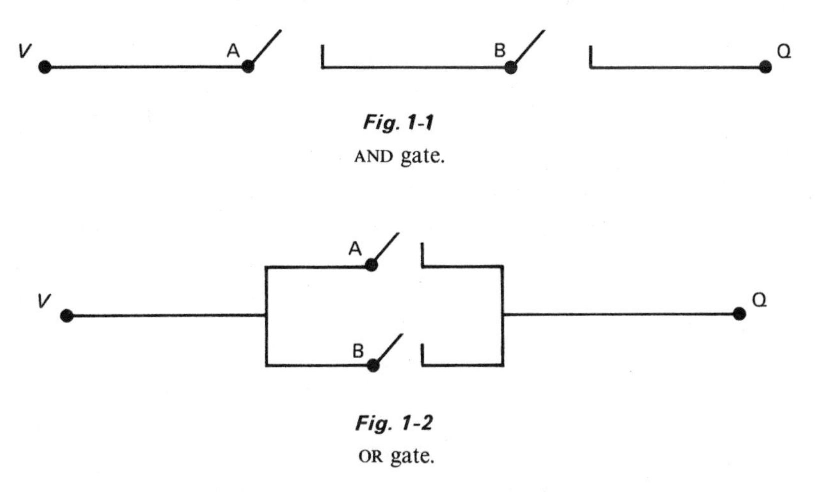

Fig. 1-1
AND gate.

Fig. 1-2
OR gate.

Table 1-1

Truth Table for AND Gate

Switch A (input 1)	Switch B (input 2)	Output (point Q)
0	0	0
0	1	0
1	0	0
1	1	1

either switch is open, then there will be no positive voltage at point Q, which would correspond to a "False" or "0" state. The type of switch setup of Fig. 1-1 corresponds to an AND gate. It is represented by the logic symbol

AND

where the wires attached to the flat side represent inputs, and the wire attached to the curved portion represents the output.

If we consider a closed switch or a higher voltage state as a True or logical "1" condition, and consider an open switch or a lower voltage state as a False or logical "0" state, we can construct a *truth table* (Table 1-1) which shows what the result will be at the output of the gate (point Q) for all combinations of switches 1 and 2.

In Fig. 1-2, note that if either switch A OR switch B is closed, the voltage applied at point V will appear at point Q. This is an OR gate and is given the logic symbol

OR

See Table 1-2 for the truth table for OR gates.

Table 1-2

Truth Table for OR Gate

Switch A (input 1)	Switch B (input 2)	Output (point Q)
0	0	0
0	1	1
1	0	1
1	1	1

Similar kinds of gates can be built in which the voltage at the output is just inverted from that expected. This is accomplished by insertion of an element called an *inverter* just prior to point Q. The logic symbol

is used to represent this element which simply serves the function of producing a logical "1" output if its input is a logical "0," and a "0" output if its input is a "1." In terms of the switches of the first two figures, this can be visualized as a third switch whose status is determined by the states of the other two. Thus, in Fig. 1-3, switch C is a relay switch which is closed when A and B are open and open when A and B are closed. Similarly, in Fig. 1-4, switch C is

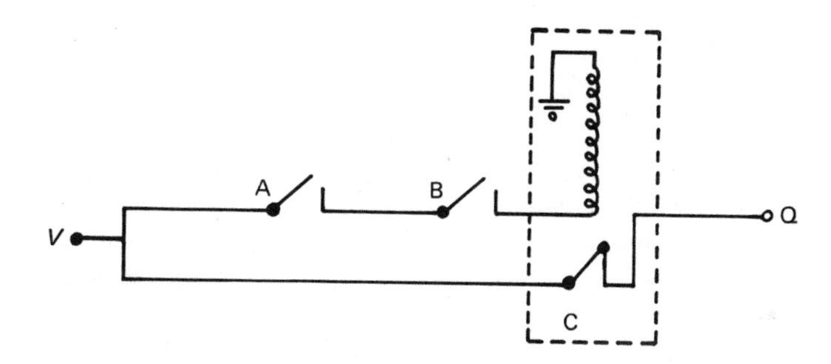

Fig. 1-3
NAND gate.

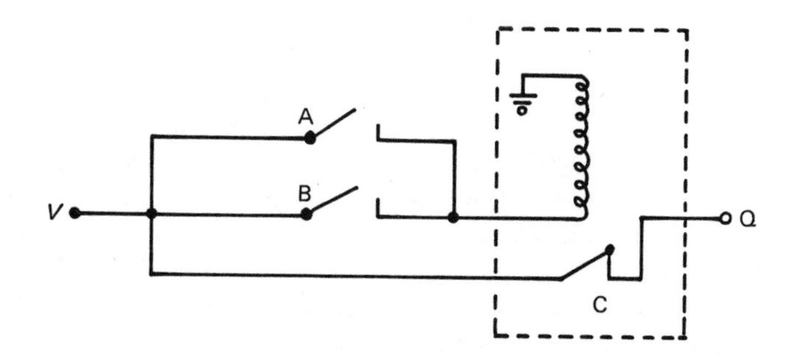

Fig. 1-4
NOR gate.

Table 1-3
Truth Table for NAND Gate

Switch A (input 1)	Switch B (input 2)	Output (point Q)
0	0	1
0	1	1
1	0	1
1	1	0

Table 1-4
Truth Table for NOR Gate

Switch A (input 1)	Switch B (input 2)	Output (point Q)
0	0	1
0	1	0
1	0	0
1	1	0

closed when both A and B are open, otherwise it is open. These are NAND and NOR gates, for "NOT AND" and "NOT OR." They are represented by the logic symbols

NAND and NOR

respectively. See Tables 1-3 and 1-4 for the truth tables for NAND and NOR gates. The actual gates, however, are not constructed from simple switches as used for the illustration above, but rather from combinations of transistors, resistors, and/or diodes which give rise to the particular logical function; the "switches" in these circuits are electronic rather than mechanical, and are controlled by applying *voltages* to the appropriate "switch" connections. Complex functions and operations such as adding, counting, etc. can be carried out by combining the "gates" in a proper fashion. It is necessary that the basic operation of these simple gates be clearly understood, if they are ultimately to be combined to implement a particular and desired function.

The various gates which we will use in these experiments utilize +5 V for one state and 0 V for the other. For our purposes, +5 V is defined as a TRUE (*logical one*) state and 0 V as a FALSE (*logical zero*) state. A problem arises regarding the precision of the voltage input required in order to obtain a particular output state. In other words, what variation in the input voltages will be "recognized" by the device as TRUE and FALSE, and where does the transition from TRUE to FALSE occur? Let us examine what happens to the output of an OR gate when one input is always held at 0 V and we vary the voltages applied to the other input. Normally, if the second input is at 0 V, the output will be 0 V; and if it is +5 V, the output will be at +5 V. As the voltage applied to this input is slowly increased, at some point the output voltage should rise until it is also at +5 V. If, on the other hand, the input voltage is at +5 V, then the output will also be at +5 V. As this voltage is slowly decreased, at some point the output voltage must go to 0 V. Fig. 1-5

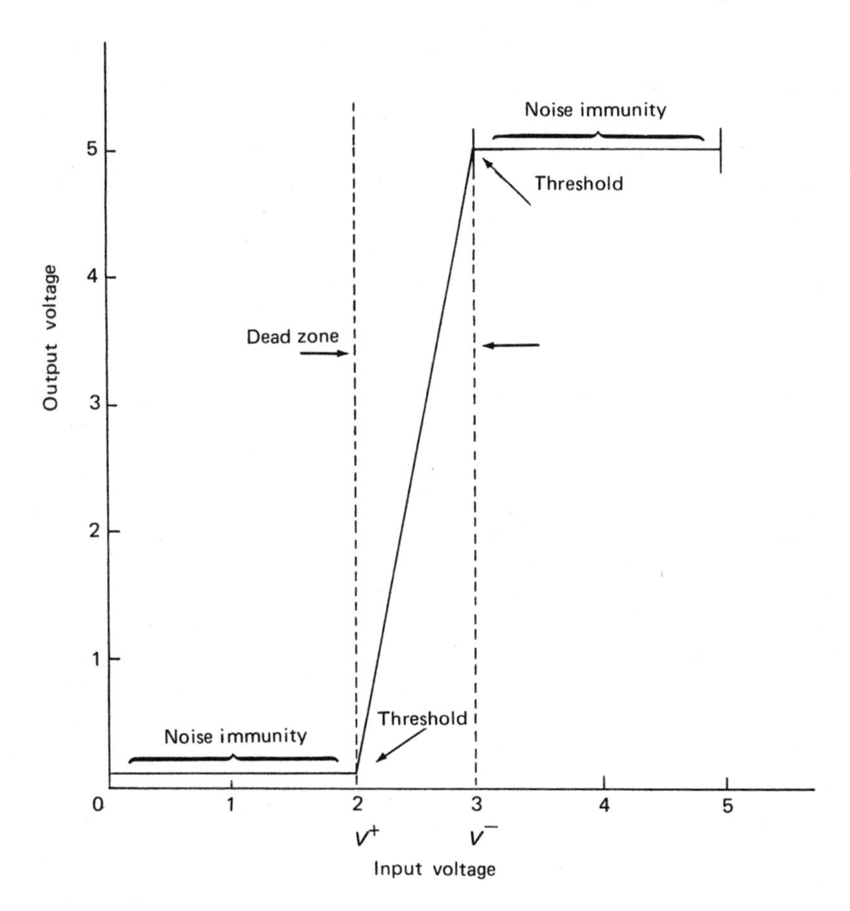

Fig. 1-5
Variation of output voltage with input voltage for a hypothetical OR gate (all other inputs grounded).

shows the approximate variation in input versus output voltage for a hypothetical OR gate. The voltage change which can occur at the input without causing a change in the output voltage is called the "noise immunity." For the hypothetical OR gate of Fig. 1-5, this immunity is about 2 V; that is, the output of the OR gate will remain at 0 V as long as the input voltage is between 0 and 2 V; correspondingly, the output will remain at +5 V if the input is between 3 and 5 V. When the input is at 0 V and begins to rise, the output does not begin its transition until the input voltage has crossed about 2 V. Similarly, if the input is at +5 V, the output will remain at +5 V until the input voltage has decreased to around 3 V. The value necessary to cause the beginning of the change in the output voltage is called the "threshold" value.

Note that the threshold values are different depending on the previous value of the input voltage. The voltage region between these two threshold values is called the "dead zone" (see Fig. 1-5) because the output observed when the voltages in these regions are applied will be neither True nor False but in between. In actual practice, the dead zone may be only a few tenths of a volt rather than the 1 V shown, and the lower threshold may lie well below 2 V. Although the discussion presented here is for an OR gate, it also applies to the other gates.

The fact that the output voltage of a gate is quite stable at one of two values (subject only to the stability of the power supply of the gate and not dependent on the stability of the input voltage) can be utilized for several purposes. For example, suppose that a signal from some device has essentially only two levels (such as 1 and 5 V), has a good bit of noise, and changes between the levels are to be used to trigger some other event. If the noisy signal is input to a gate such as an OR gate, then the output signal will be quite stable at 0 or 5 V. Fig. 1-6 shows the signals. Thus, even if the input signal is noisy, so long as it is at about 0 V or 5 V and the noise does not exceed 2 V, the output voltage will be 0 and 5 V $\pm$ the stability of its power supply. The input signal is cleaned up and appears very stable at the output.

Occasionally, a signal will have more than 2 V of noise. An additional feature of logic gates is that if the output is fed back to the input through a

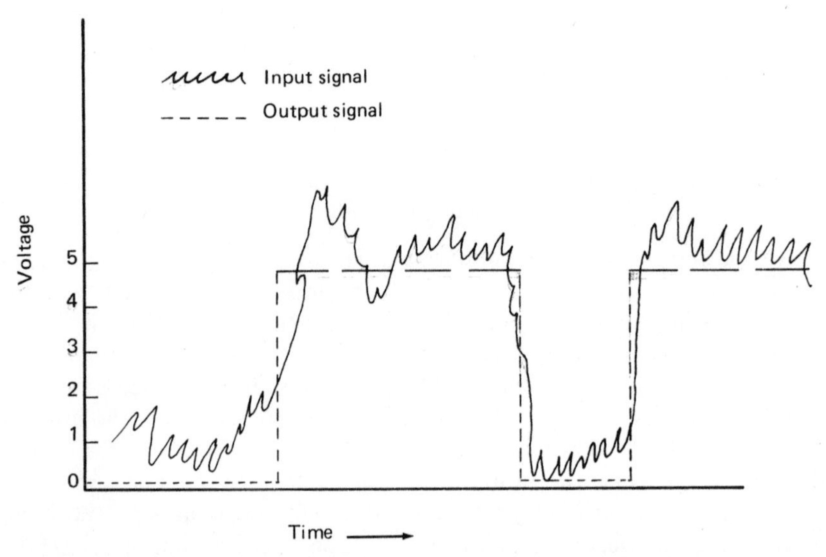

Fig. 1-6
Result of inputting a noisy signal to an OR gate.

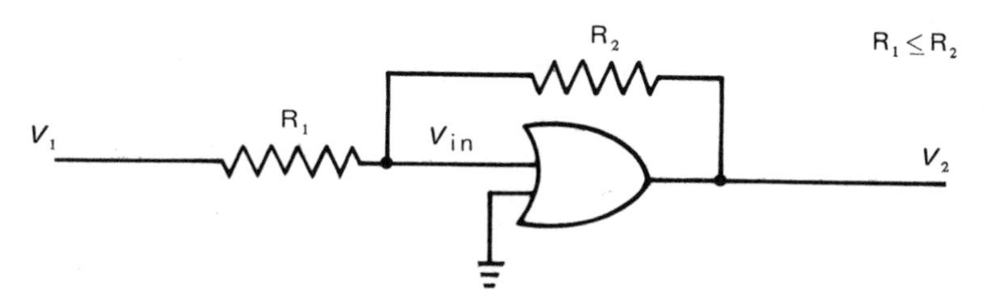

Fig. 1-7
A Schmitt trigger.

voltage-divider network, it is possible to shift the thresholds to minimize the dead zone and maximize the noise immunity for such signals. Consider, for example, the circuit shown in Fig. 1-7. The voltage at the gate input, V_{in}, is given by the expression

$$V_{in} = \frac{(R_1 V_2 + R_2 V_1)}{(R_1 + R_2)} \tag{1}$$

When V_1 and V_2 are initially at zero, and V_1 begins to rise, V_{in} will be *less* than V_1, due to the low voltage fed back through R_2. If the threshold voltage necessary to trigger a zero-to-one transition is V_+ (see Fig. 1-5), then V_1 must be at least $(R_1/R_2)V_+$ *above* V_+ in order to cause a transition (assuming V_2 is zero). The noise immunity is thus increased by $(R_1/R_2)V_+$.

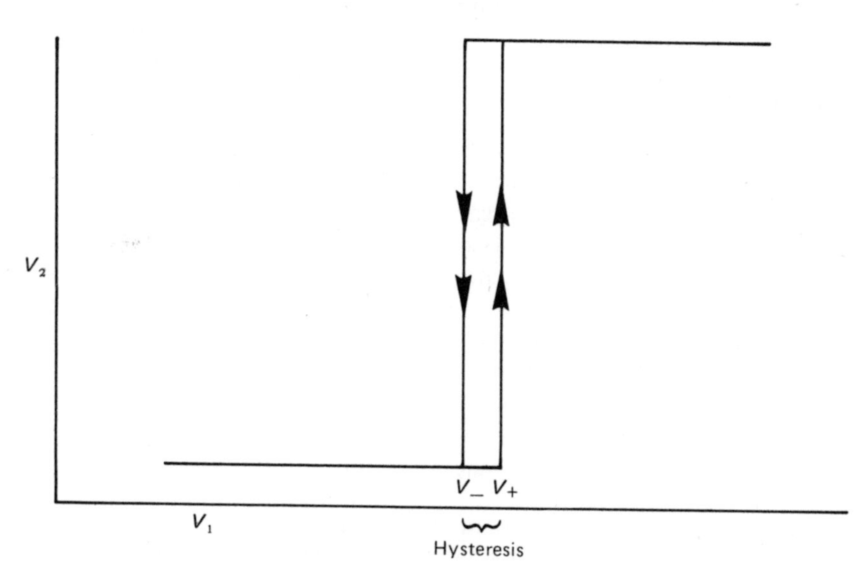

Fig. 1-8
Hysteresis curve of a Schmitt trigger.

Now consider the case where V_1 and V_2 are initially 5 V, and V_1 begins to drop. In this case, V_{in} will be higher than V_1 due to the contribution of V_2. Then V_1 must be $(R_1/R_2)(V_2 - V_-)$ *below* V_- before a transition occurs, and the noise immunity is again increased. In actual application, when V_+ and V_- lie within 0.2–0.3 V of each other, the voltage necessary at V_1 to trigger a zero-to-one transition may be *greater* than the voltage necessary to trigger a one-to-zero transition. Although this may sound like a contradiction in terms, in practice it works quite simply, as is shown in Fig. 1-8. When a signal V_1 rises above V_+, then V_2 changes from zero to one. Should V_1 now drop below V_+, but remain above V_-, V_2 will remain unchanged. Only when V_1 drops below V_- will V_2 change; similarly, V_2 will not change from zero to one unless V_1 is greater than V_+. A circuit with those properties is called a *Schmitt trigger.* The difference between the overlapping thresholds is called the *hysteresis* or *backlash level.*

EXPERIMENTAL

This experiment is designed to introduce you to the various gates employed in computers and in the interfacing of an experiment to a computer. You will observe the static conditions of AND, OR, NAND, and NOR gates. You will also observe the effects on the output of each gate of varying input sources. After becoming familiar with each type of gate, you will see how these gates may be utilized in building a simple Schmitt trigger, observing input and output signals as well as the effects of varying certain adjustable parameters.

Equipment Needed

1 EL Digi-designer (see Appendix C)
1 5-V variable power supply
1 Quad dual-input AND gate package (SN 7408)*†
1 Quad dual-input NAND gate package (SN 7400)†
1 Quad dual-input OR gate package (Signetics SP 384A)†
1 Quad dual-input NOR gate package (SN 7402)†
1 Sine-wave generator
1 Volt–ohm multimeter (VOM)
1 Oscilloscope (dual trace works best)
2 Variable resistance boxes (15 Ω–10 MΩ)

*TTL 7400 series part numbers will usually be specified in this manual. The instructor may specify alternate equivalent gates.

†Integrated circuits are packaged in several forms. The most common one used is the dual-in-line J or N (Signetics A, B, or N) package where one or more gates or functions are contained on one chip. The pins point downward. The chips suggested above are N packages.

Procedure

Be sure to check the data sheets (to be supplied by your instructor) for each of the logic chips very carefully. Refer to Appendix G for an explanation of how to read a schematic. Identify each of the pins associated with each gate of the chip. *Avoid connecting together the outputs of any two gates as this may destroy both gates.* Although an open connection on any input usually acts the same as a high-voltage input (logical 1), this is *not true in all cases*; in particular, this is not the case for the OR gate package recommended above. It is poor wiring practice to leave unused inputs open. Rather, they should be wired to inputs which are used. For each of the four gates AND, NAND, OR, and NOR, perform the following steps:

A. Logic Functions and Gates

1. Consult the data sheet for the integrated circuit chip used and wire the appropriate pins to the +5 volt power supply (V_{cc}) and ground (GND). This will provide the power for operation of the circuit.
2. Using the Digi-designer logic switches, apply all possible voltage combinations to the inputs of *one* of the four gates (i.e., +5 V and 0 V, 0 V and 0 V, etc.). Use a Digi-designer indicator lamp or a VOM to monitor the output state for each combination of inputs and, hence, to determine the truth table for the gate. Record your results and write the truth table for the chip.
3. Connect the negative (ground) lead of a variable power supply to the ground lead of the 5-V power supply or Digi-designer. Connect one gate input to the ground and, using the variable power supply, apply in a stepwise fashion an increasing voltage from 0 to +5 V to the remaining input (a step size of about 0.2 V works well here). Monitor the gate output voltage with either a volt–ohm multimeter (VOM) or an oscilloscope. If a transition occurs, be sure to measure a sufficient number of points in the "dead zone." *Note*: Some chips, particularly the SN 7408, tend to oscillate near the transition point. You may skip Sec. 3, 4, 5, and 7 for chips which oscillate.
4. Next, switch one input to the +5-V source, and to the other input apply an increasing voltage while monitoring the output as in step C.
5. Plot output voltage versus input voltage for those ranges in which transitions occur.
6. Compare the truth tables for the AND and OR gates. Also compare the truth tables for the NAND and NOR gates.
7. *Option*: If an ammeter is available, monitor and record the input and output currents for one or two of the gates.

B. Schmitt Trigger

1. Construct the circuit shown in Fig. 1-7, using variable resistance boxes for the resistors. Set R_1 and R_2 to 1000 and 3300 Ω, respectively. Be sure to power the OR chip according to the data sheet.
2. Using a sine-wave generator, apply a sine wave to V_{in}. Monitor V_{in} and V_{out} simultaneously on a dual-trace oscilloscope; trigger *both* sweeps on V_{in}. Sketch the observed curves.
3. Vary R_1 and R_2 independently. Monitor and record any changes in V_{out}. What happens when R_1 is greater than R_2?

Experiment 2

Complex Logic Functions from Simple Gates—Part 1*

INVERTER

In addition to the simple AND, NAND, OR, and NOR gates you are now familiar with, there are several other similar functions which can easily be generated by the proper combination and sequence of the AND, NAND, etc., gates. The first of these is a simple invert function whose symbol is

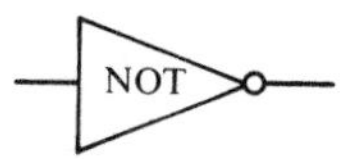

This function can be generated from any negative (NAND or NOR) gate by connecting all but one of the inputs to a fixed voltage (+5 for a NAND, 0 for a NOR), while the last input is connected to the signal to be inverted. The output is, therefore, always the opposite of the input state.

EXCLUSIVE OR

The second gate which can easily be generated by a combination of NOR, OR, and NOT gates is the Exclusive OR (XOR) whose symbol is

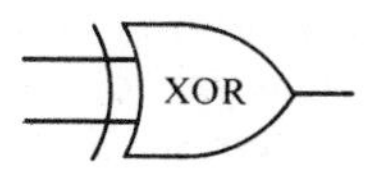

(See Fig. 2-1). The XOR gate can also be constructed from AND, OR, and NOT

*Inverter, exclusive OR, RS flip-flop, clocked RS flip-flop, RS and JK master–slave flip-flops, and buffered switches.

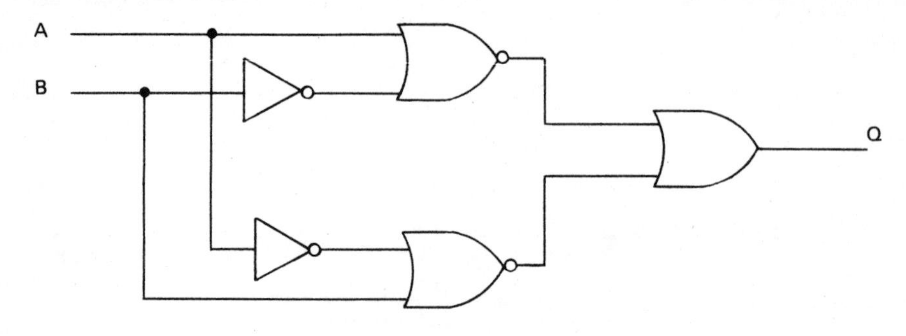

Fig. 2-1
Exclusive OR.

Table 2-1
XOR Truth Table

A	B	Q (output)
0	0	0
0	1	1
1	0	1
1	1	0

gates. This gate has the unique quality that, when either of the two inputs (*but NOT both*) is true (logical 1), the output of the gate is true. The truth table for the XOR gate is shown in Table 2-1.

RS FLIP-FLOP

Now that the basic operation of the simple gates has been considered, you may proceed to the building of more complex functions by combining these gates in particular patterns so that the output will be both known and predictable. The simplest of the more complex functions is an RS (for Reset–Set) flip-flop. This is a device which has two stable conditions. It will remain in one of these two states until it is forced out of it by an appropriate change at the input. Thus, the flip-flop acts as a binary memory device. The RS flip-flop is easily constructed from a pair of 2-input NAND gates as shown in Figure 2-2. The truth table for this circuit is shown in Table 2-2.

Note that the outputs of the two gates are labeled Q and $\overline{Q}$ respectively. The symbol $\overline{Q}$ indicates that this output should be the logical complement of Q. The inputs are labeled S and R. As long as S and R are not simultaneously at ground (logical 0) the outputs at Q and $\overline{Q}$ are predictable and complementary.

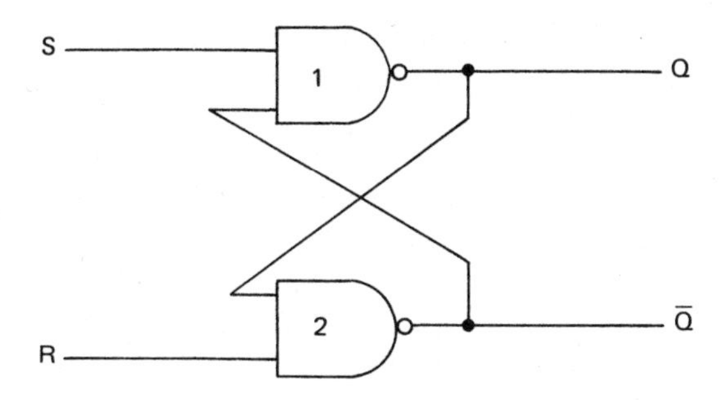

Fig. 2-2
RS flip-flop from cross-coupled NAND gates.

When S and R are both logical zero, Q and $\bar{Q}$ are both logical one, violating the initial requirement that they be complementary! Moreover, should S and R return to logical one simultaneously, *either* Q *or* $\bar{Q}$ will become logical zero, but *we cannot predict which will change.* For this reason, the case where S and R are both logical zero is considered undefined or indeterminate. All other states are defined, as Table 2-2 shows. The standard symbol for the simplest of RS flip-flops is shown in Fig. 2-3.

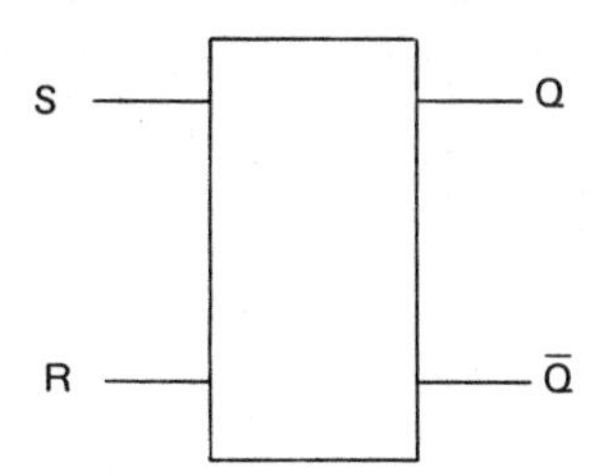

Fig. 2-3
RS flip-flop symbol.

Table 2-2
Truth Table as RS Flip-Flop

S	R	Q	$\bar{Q}$
0	0	Undefined	
0	1	1	0
1	0	0	1
1	1	(Unchanged from previous state)	

The utility of the simple RS flip-flop is quite limited and it must be modified to be useful. This can be demonstrated if one considers the effect of connecting a pulse train of 1, 0, 1, 0, 1, . . . to the *S* input, while *R* is held at +5 V. If the *Q* output is initially at logical 0, then the first time the pulse level goes from 1 to 0, the *Q* output will change from 0 to 1. However, any additional pulses will not cause a change at *Q* (see truth table). An analogous situation occurs if the pulse train is connected only to the reset (or clear) input while *S* is held at +5 V. Should both inputs be connected together, then the output will be undefined when the pulse train undergoes the 0-to-1 transition. Thus, the *RS* flip-flop can be used to remember one and only one event or change.

CLOCKED RS FLIP-FLOP

A more useful flip-flop can be obtained by running the *S* and *R* inputs through two additional 2-input NAND gates with the outputs going to the normal *S* and *R* inputs of a simple RS flip-flop. Fig. 2-4 shows a diagram of such an arrangement. This particular logic circuit is called a gated or clocked RS flip-flop. It has the advantage that, regardless of the state of the *R* or *S* inputs, the output of the flip-flop cannot change until the *C* input is made logical 1 (+5 V); that is, the outputs of NAND gates 1 and 2 will always be logical 1 when *C* is at logical 0. Note also that, for nonzero *C*, gates 1 and 2 act as inverters; that is, *Q* is set when *S* is one (and *C* is one) rather than when *S* is zero. This is opposite to the earlier case. The complete truth table for the clocked RS flip-flop more clearly shows the results of various combinations of *S*, *R* and *C* inputs (Table 2-3). This flip-flop also suffers from some of the

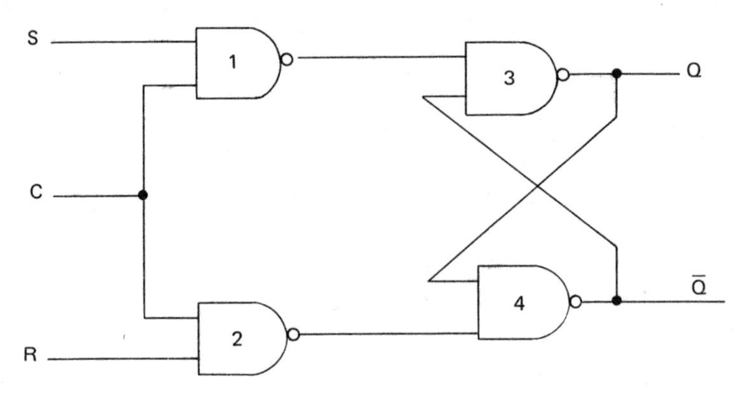

Fig. 2-4
Clocked RS flip-flop.

Table 2-3

Truth Table Clocked RS Flip-Flop

S	R	C	Q	$\bar{Q}$
0	0	0		
0	1	0		
1	0	0	No change	No change
1	1	0		
0	0	1		
0	1	1	0	1
1	0	1	1	0
1	1	1	Unknown or not predictable	

same shortcomings as the simple RS flip-flop. With further modifications, however, it can be used to perform counting operations (see following section).

MASTER–SLAVE FLIP-FLOPS

A more complicated type of flip-flop can be constructed by linking two of the simpler RS flip-flops in a way such that the second or "slave" flip-flop follows the first or "master" flip-flop when the clock pulse allows it. In this way, the inputs may be *buffered* from the output, ensuring that no change of the input conditions will affect the output until the end of the clock pulse.

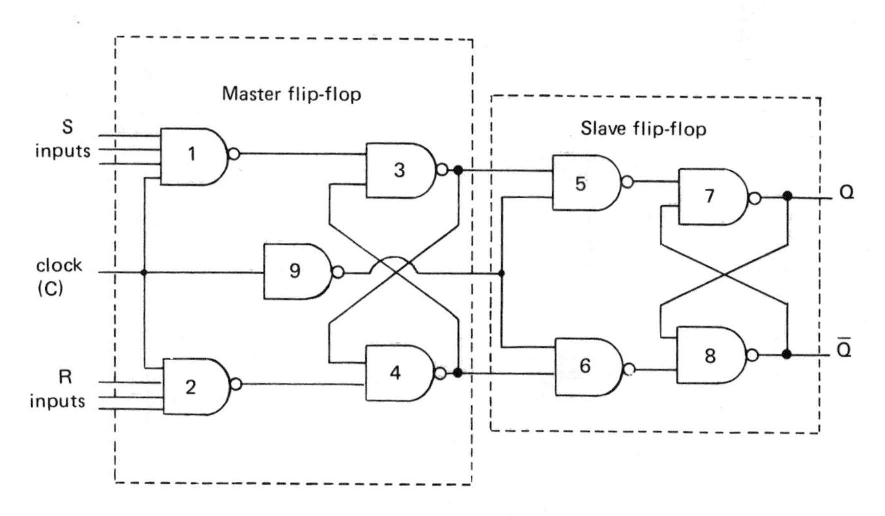

Fig. 2-5

RS master–slave flip-flop from NAND gates.

Table 2-4

Truth Table for RS Master–Slave Flip-Flop*

		C			Master			Slave		
S	R	t	$t+1$	$t+2$	t	$t+1$	$t+2$	t	$t+1$	$t+2$
0	0	0	1	0	Q_t	Q_t	Q_t	Q_t	Q_t	Q_t
0	1	0	1	0	Q_t	0	0	Q_t	Q_t	0
1	0	0	1	0	Q_t	1	1	Q_t	Q_t	1
1	1	0	1	0	Q_t	Undefined		Q_t	Q_t	Undefined

*t is any particular time; $t + 1$ is an instant later, when the clock signal has changed logic states; $t + 2$ is another instant later, when the clock signal has returned to its original level; Q_t is the logic state of Q output at time t.

Such an arrangement allows the output (Q or $\bar{Q}$) to be "read" by another device such as a computer while ensuring that the outputs remain stable until that operation is complete (provided the reading device can control the clock pulse).

Figure 2-5 is a detailed wiring diagram for an RS master–slave flip-flop. The truth table for this circuit is shown in Table 2-4. The quantity Q_t represents the state of the master flip-flop Q at time t. The notation $t + 1$ indicates a short time after time t, when the clock signal has just changed logic levels;

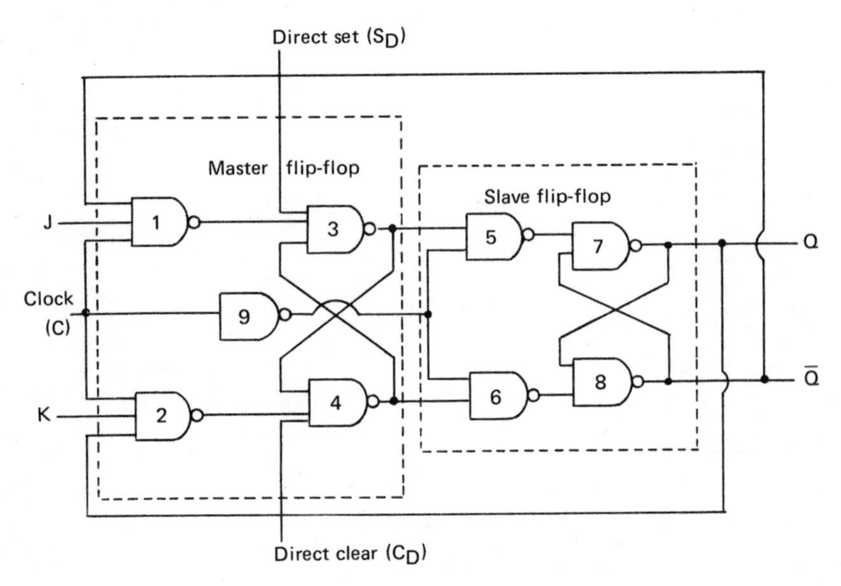

Fig. 2-6

JK master–slave flip-flop from NAND gates.

Table 2-5

Truth Table for JK Master–Slave Flip-Flop

		C			Master			Slave		
J	K	t	$t+1$	$t+2$	t	$t+1$	$t+2$	t	$t+1$	$t+2$
0	0	0	1	0	Q_t	Q_t	Q_t	Q_t	Q_t	Q_t
0	1	0	1	0	Q_t	0	0	Q_t	Q_t	0
1	0	0	1	0	Q_t	1	1	Q_t	Q_t	1
1	1	0	1	0	Q_t	$\bar{Q}_t$	$\bar{Q}_t$	Q_t	Q_t	$\bar{Q}_t$

the notation $t + 2$ then indicates a yet later time when the clock signal has just returned to its original state. Note that while the master flip-flop is active for clock values of logical one, the slave is inactive, and conversely. Thus, if the master changes at $t + 1$ when the clock signal is logically true, this change is not transmitted to the slave until $t + 2$, when the clock signal drops to 0 V. But at that point, further changes in R and S will not affect the master and, hence, cannot change the slave output. Unfortunately, this more complex circuit still suffers from the occurrence of an undefined or indeterminate state.

Fortunately, it is possible to devise a flip-flop (also constructed from simple NAND gates) which has *no* undefined states. By connecting the output lines of the slave flip-flop back to the input lines of the master flip-flop, a new type called a JK master–slave flip-flop can be built. Since there are no undefined states, this particular type of flip-flop is far more useful than the simpler ones discussed earlier.

Figure 2-6 shows the diagram for construction of a JK master–slave flip-flop using NAND gates. Table 2-5 shows the truth table for this master–slave flip-flop. The notation is that of Table 2-4, except inputs S and R have been replaced by J and K, respectively. The truth table is similar to that of the RS master–slave, except that *all* states are now defined. When both inputs are logically true, and the clock is pulsed, the flip-flop will "toggle" or change to the complement of its previous state. This feature is especially useful in counting applications, and will be discussed in greater detail in a later experiment. As in the case of the RS master–slave flip-flop, at $t + 2$ the slave assumes the state of the master at $t + 1$.

Diagrams which show the condition of the various inputs and outputs as a function of time are useful to illustrate the point. These are called "timing diagrams" and are used to keep track of the time relationships of events in complex systems. Figure 2-7 shows such a diagram for the JK flip-flop of Fig. 2-6. Assume that the J and K inputs are both held at logical 1. Note that the slave output assumes the value of the master output when the

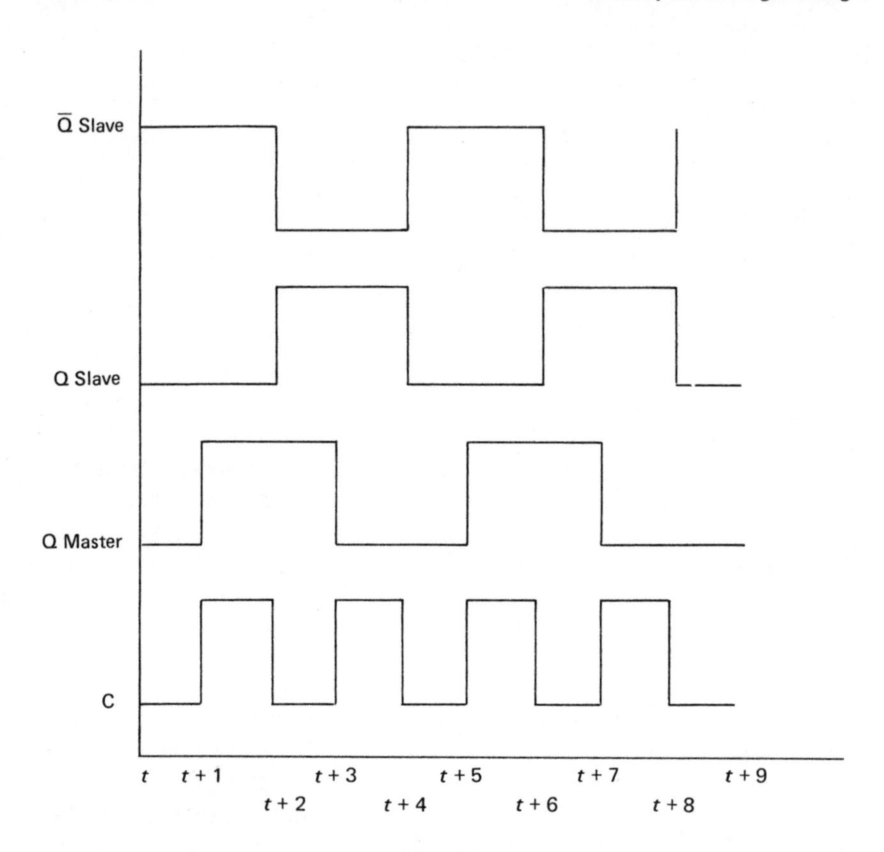

Fig. 2-7
Timing diagram for JK master–slave flip-flop (J and K inputs logical 1).

clock input goes from 1 to 0. Also, note that the master cannot change when the clock input is 0. Because the only output normally accessible for commercial master–slave flip-flops is the slave output, it appears that the flip-flop can only change state during the clock *transition* from 1 to 0. Thus, this is often referred to as "trailing-edge-triggered" logic. If a clock transition from 0 to 1 allows the state change, it is called "leading-edge-triggered" logic.

The master–slave JK flip-flop is very useful in constructing various counters and registers used in computer interfacing. In actual practice, the various flip-flops described here are not usually constructed by the user from simple gates as shown, but rather are bought as integrated-circuit packages. The inputs at which the tiebacks from Q and $\bar{Q}$ exist, as well as other internal connections, are seldom shown on the data sheets accompanying the commercial IC (integrated circuit) package.

BUFFERED SWITCH

If one were to construct the various flip-flops from NAND gates and subsequently use simple toggle switches to control the logic, one might not get results which would be either meaningful or consistent with the truth tables shown. The reason for this is the fact that spring-loaded switch contacts bounce when closed. The result is that when the switch is thrown, the resulting output voltage may oscillate, giving rise to an apparent several-fold change in logic state for one closing of the switch. This problem can be overcome by use of a "buffered switch" (that is, a switch which will give one and only one change at its output each time it is closed). This buffering action can be obtained by inserting a RS flip-flop between the switch and its output. (Refer to Table 2-2 for the simple RS flip-flop.) Figure 2-8 is the wiring diagram for a buffered switch. When the switch is thrown from 1 to 2, the S input goes from 0 to 1 (since open inputs are "true"), and R goes from 1 to 0. The truth table for the RS flip-flop shows that the Q output is initially a logical 1, and changes to a logical 0 as the 2 position is first reached. If the contacts bounce, both inputs will be in a 1 state momentarily, but no change in the output of Q or $\bar{Q}$ occurs, so long as the switch does not bounce all the way back to the 1 position.

EXPERIMENTAL

The purpose of this experiment is to construct the various logic functions and flip-flops so that you may familiarize yourself sufficiently with them to build other useful devices such as counters, registers, or data latches.

Equipment Needed

1 EL Digi-designer
2 Triple 3-input NAND gates (SN 7410)
1 Quad 2-input NOR gate package (SN 7402)
1 Quad 2-input OR gate package (Signetics SP 384A)
6 Quad 2-input NAND gate packages (SN 7400)
1 Hex inverter package (SN 7404)
1 Pulse generator (0 to 5 V), variable frequency
1 Storage oscilloscope

Procedure

Caution: Remember to consult the data sheets for each device. *Connection of the outputs of two TTL gates may destroy both.*

Power each package according to the data sheet, noting which pins are connected to the appropriate sources or states.

It is strongly suggested that you draw a diagram of each package, numbering each gate and drawing in the pin number connections you are going to make.

A. Inverter

1. Construct an inverter from a single NAND gate by following the circuit below. Verify that Q is the inverse of A, regardless of A.

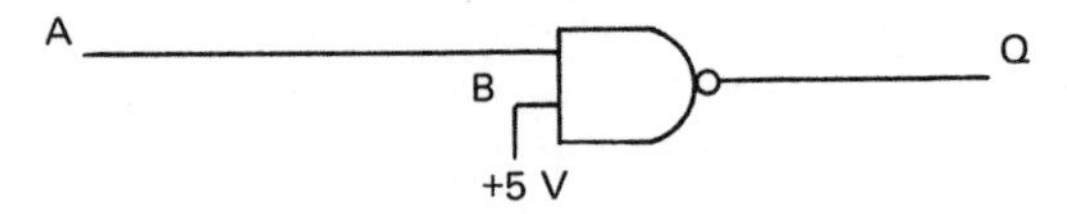

2. Replace the NAND gate package with a NOR gate package. Connect the B input to *ground*. Verify that this circuit, too, operates as an inverter.
3. Wire the inverter package as shown on its data sheet. Verify that this performs in the same manner as the NAND or NOR gate wired as an inverter.

B. Exclusive OR (XOR)

1. Wire the diagram as shown in Fig. 2-1.
2. Verify the truth table for the XOR function.

C. RS Flip-Flop

1. Wire the diagram as shown in Fig. 2-2. Connect the Q and $\bar{Q}$ outputs to separate lights in order to monitor their states. Connect the S and R inputs to logic switches, initially set at 5 V.
2. Switch the S input to ground and note the states of Q and $\bar{Q}$.
3. Now switch S back to +5 V and switch R to ground. Note the effect on Q and $\bar{Q}$.
4. Switch both S and R inputs to ground. Do this several times, and note if the initial value of Q has any effect.
5. From your observations construct a truth table and compare it with that shown for the RS flip-flop. Can you successfully predict the state of Q and $\bar{Q}$ for a variety of states of S and R?

D. Gated or Clocked RS Flip-Flop

1. Construct the circuit shown in Fig. 2-4 using NAND gates. Connect C to a logic switch, originally set to ground. Monitor the Q and $\bar{Q}$ outputs with lights.

2. Switch S to ground. Does the state of Q change?
3. Switch S back to +5 V, and switch R to ground. Does Q change state?
4. Now switch both to ground. Is any change observed in Q? (Do this several times.)
5. Now switch C to +5 V and repeat steps 2 to 4.
6. Compare the data collected with the truth table for the clocked RS flip-flop (Table 2-3.)

E. Buffered Switches

1. Wire the circuit shown below. Connect the storage oscilloscope and set it at a rapid scan of 100 μsec/div., vertical scale about 1 V/cm, external trigger. Alternately, a storage oscilloscope with continuous scan can be used; however, the patterns obtained when the switch is opened and closed several times will not be synchronized.

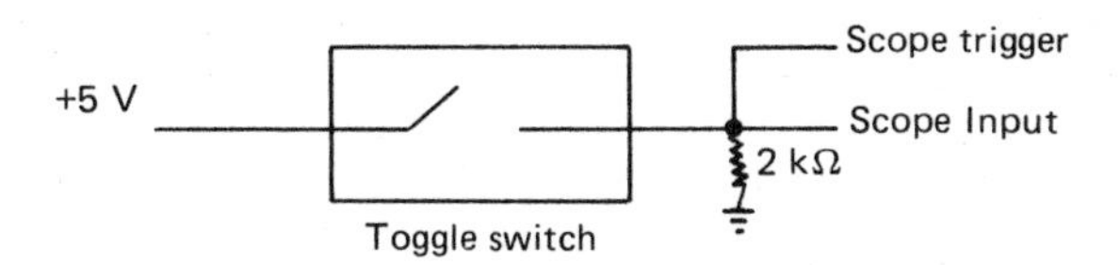

2. Close and open the toggle switch several times, and try to observe any "bounce" or fluctuation as the switch is opened and closed.
3. Construct the circuit shown in Fig. 2-8. Attach the oscilloscope to the Q output. Open and close the switch several times. The voltage signal on the scope should be very stable with little or no "bounce." Leave this switch wired for use in the next section.

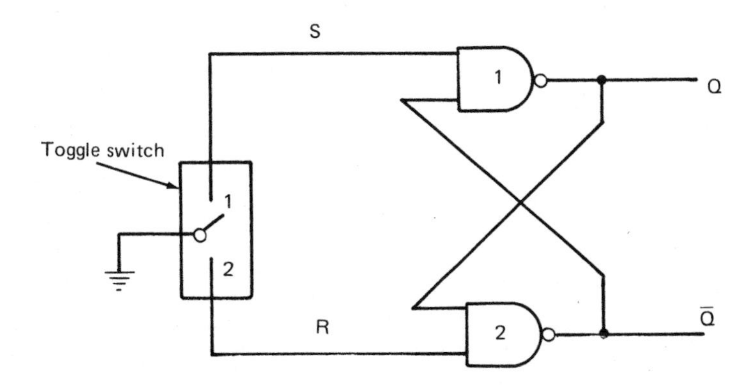

Fig. 2-8
Buffered switch.

F. JK Master–Slave Flip-Flop

1. Wire the circuit as shown in Fig. 2-6. Monitor the Q output of both the master and slave with lights. Use a buffered switch at the C input.
2. Verify the truth table (Table 2-5) for this flip-flop.
3. Observe the timing sequence of your flip-flop and compare it with the timing diagram of Fig. 2-7.

Experiment 3

Complex Logic Functions From Simple Gates— Part 2*

COUNTERS

In Experiment 2, you learned how to construct various flip-flops and other logic functions. In this section, you will learn to combine these simple devices so that they may be used to perform even more complex functions. Devices such as counters, registers, and data latches will be built.

Recall that a flip-flop is a device which has two states, True (1) and False (0), and that it remains in a given state until it is forced out by some change in the input state. Remember too that, with JK inputs held at +5 V, a JK flip-flop "toggles" or changes to the opposite output state every time a clock pulse (1 → 0) occurs. If several of these JK flip-flops are connected together with each Q output going to the C input of the succeeding JK flip-flop, you have a simple binary counter. (All registers and counters discussed here are constructed using master–slave JK flip-flops.)

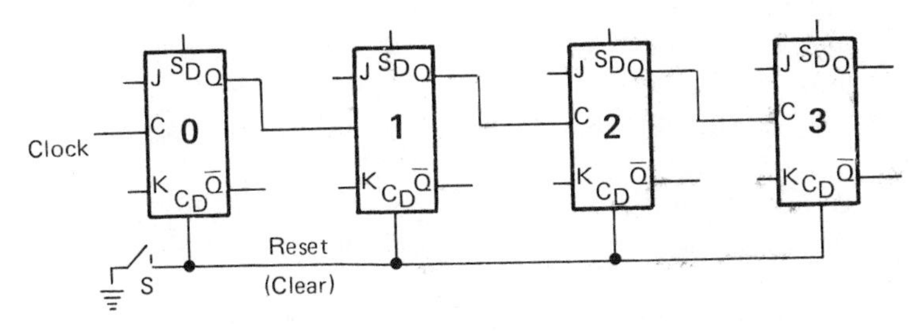

Fig. 3-1
Asynchronous binary counter.

*Assorted counters, data latches, shift registers, and one-shots.

Table 3-1

States of Flip-Flops for N Pulses

Pulse number	Q_3 FF3	Q_2 FF2	Q_1 FF1	Q_0 FF0
0	0	0	0	0
1	0	0	0	1
2	0	0	1	0
3	0	0	1	1
4	0	1	0	0
5	0	1	0	1
6	0	1	1	0
7	0	1	1	1
8	1	0	0	0
9	1	0	0	1
10	1	0	1	0
11	1	0	1	1
12	1	1	0	0
13	1	1	0	1
14	1	1	1	0
15	1	1	1	1
16	0	0	0	0

Figure 3-1 shows the set-up for a four-stage binary counter. Note that the detailed diagrams for each JK flip-flop, as previously given in Fig. 2-5, are not shown, and an abbreviated symbol is used. Each of the flip-flops (FF) is numbered (0 to 3). If the J and K inputs are left at logical 1 (remember that an open input is usually logical one), whenever C changes from 1 to 0, the output of that particular flip-flop (FF) changes or "toggles." Consider what will happen if all flip-flops are initially cleared by grounding the C_D inputs, and the following sequence occurs. A pulse ($0 \rightarrow 1 \rightarrow 0$) is applied to the C input, C_0, of FF0. The Q_0 output will change state, i.e., to a 1. On the next pulse, Q_0 will change to 0, which in turn causes the C_1 input to go from a logical 1 to 0, and Q_1 to change to a 1. On the third pulse at C_0, Q_0 will go to a 1, but Q_1 remains the same. On the fourth pulse, both Q_0 and Q_1 go to 0, and Q_2 goes to 1. Table 3-1 shows the effect of a series of pulses on the Q outputs of the various flip-flops. (*Note*: the order of the flip-flops has been reversed from that of Fig. 3-1 to facilitate the following discussion of the binary number system.)

Figure 3-2 shows the timing diagram for this counter. Note that the outputs of the flip-flops do not change until the clock goes from 1 to 0, as we are using trailing-edge-triggered JK master–slave flip-flops. Note also that Q_0 changes on every pulse, Q_1 on every second pulse, Q_2 on every fourth pulse, and Q_3 on every eighth pulse. This particular "counter" is capable of

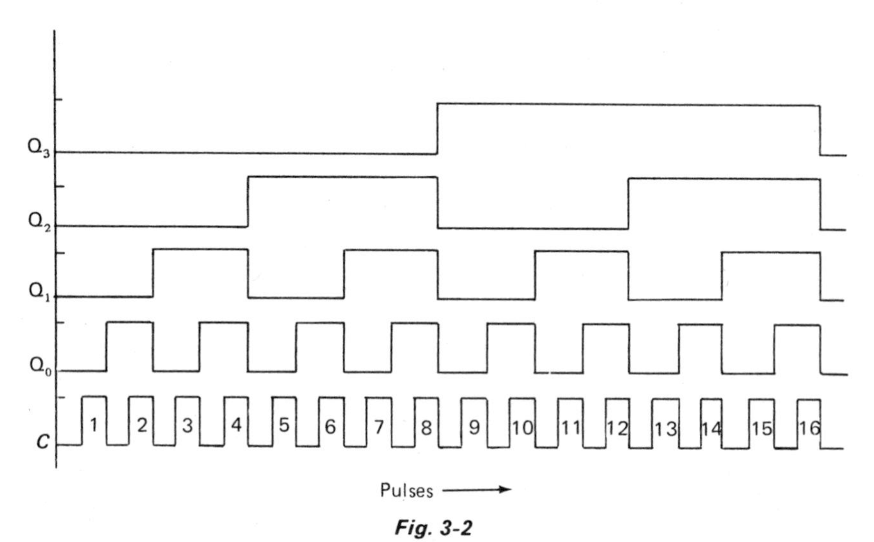

Fig. 3-2

Timing diagram for 4-bit binary counter.

counting from 0 to 15 (on the 16th pulse, all flip-flops reset to 0 and the process repeats). A count of the number of pulses can be determined by adding up the values of the flip-flops multiplied by the weight of each. Suppose we look at the following state of the flip-flops:

Q_3	Q_2	Q_1	Q_0
1	1	0	1

The value that each flip-flop is counting by can be computed by successively assigning a number, n, to each FF, beginning with the one nearest the incoming pulse. Each FF counts by 2^n, where $n = 0, 1, 2, 3, \ldots$; i.e., the nth FF changes state after every 2^n pulses. The actual number is then the sum of all flip-flop values. This binary counting system can be compared with the decimal system with which you are quite familiar. The number $321_{10} = 3 \times 10^2 + 2 \times 10^1 + 1 \times 10^0 = 300 + 20 + 1$. Similarly, the binary number $1101 = 1 \cdot 2^3 + 1 \cdot 2^2 + 0 \cdot 2^1 + 1 \cdot 2^0$, which is $8 + 4 + 1$, or 13 in the decimal system.

$$\begin{aligned} \text{FF3} &= 8\,(1 \cdot 2^3) \\ \text{FF2} &= 4\,(1 \cdot 2^2) \\ \text{FF1} &= 0\,(0 \cdot 2^1) \\ \text{FF0} &= 1\,(1 \cdot 2^0) \\ &\overline{13} \end{aligned}$$

Comparison with Table 3-1 indicates that this is, indeed, the pattern observed after 13 clock pulses have been entered into the counter. Thus, the binary number system can represent any number by the appropriate combination of ones and zeros. Obviously, the length of the combination of ones and zeros will have to be much longer than if it were expressed as a decimal number.

SYNCHRONOUS COUNTERS

The type of counter in Fig. 3-1 is an asynchronous or "ripple-carry" counter. That is, the higher-order flip-flops cannot respond to a given pulse until all lower order FF's have. Thus, the pulse which is input at C_0 will cause the flip-flops to change state sequentially as it "ripples" through the succession of flip-flops. No process occurs instantaneously; therefore, if the series of flip-flops is a long one and/or the input pulses are very frequent, a second (or third or fourth) pulse might be input to the counter before the counter has completed counting or changing as a result of preceding pulse(s).

By rearranging the connections of the various flip-flops and adding some additional gates, the entire counter can be made to change at the same time, or synchronously. Figure 3-3 shows the same series of four flip-flops, modified by the addition of two gates, so that all flip-flops change at the same time. In this configuration, FF1 cannot toggle until Q_0 is 1 *and* a clock pulse occurs. FF2 cannot toggle until both Q_0 and Q_1 are in the 1 state and a clock pulse occurs. Since all changes in state occur simultaneously, this is called a synchronous binary counter.

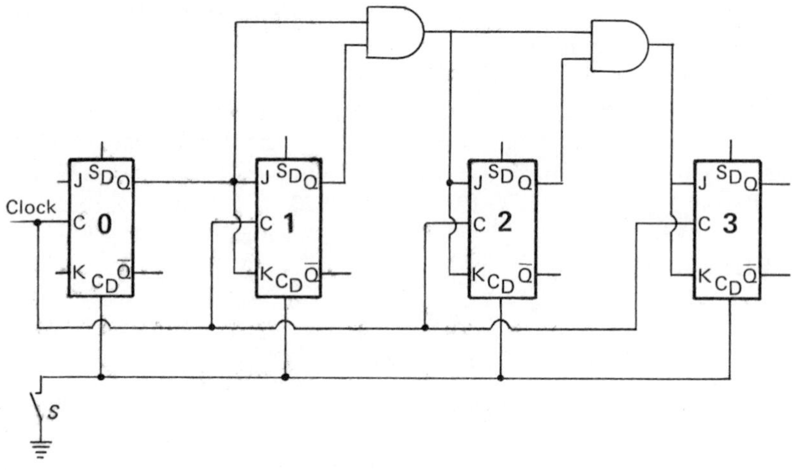

Fig. 3-3
Synchronous binary counter.

Table 3-2

Status of Binary Counter Flip-Flop

Count	FF3	FF2	FF1	FF0
8	1	0	0	0
9	1	0	0	1
10	1	0	1	0
11	1	0	1	1

This synchronous counter can be made to count up to any number between 0 and 15 by suitable gating of the outputs. For example, suppose that we want to count only to 10 and not to 16, as the counters in Figs. 3-1 and 3-3 will do. Careful examination of the state of each flip-flop as it goes from 9 to 10 to 11 gives some insight into how you can apply gates to achieve the desired end (see Table 3-2). To count to 10, we want the flip-flops to be set to zero on the tenth pulse. Q_1 and Q_2 are already at zero, so they should be held there on the tenth pulse, while Q_0 and Q_3 go to zero. This can be accomplished relatively easily by the use of some additional AND gates as shown in Fig. 3-4. First, remember that we want Q_1 to change only when Q_3 is at 0 (i.e., <8 count). To accomplish this, connect the $\bar{Q}$ output of FF3 and the Q output of FF0 to the inputs of AND gate No. 1. Now, Q_1 can only change when Q_3 is 0, Q_0 is 1, and a clock pulse occurs. In a similar fashion, using AND gate No. 2, we can "trick" FF3 into "believing" that it is seeing a 16 count, and it will toggle or change its Q output to 0 on the tenth pulse. The result is that the counter will now count to 9 and on the tenth pulse go to all zeros.

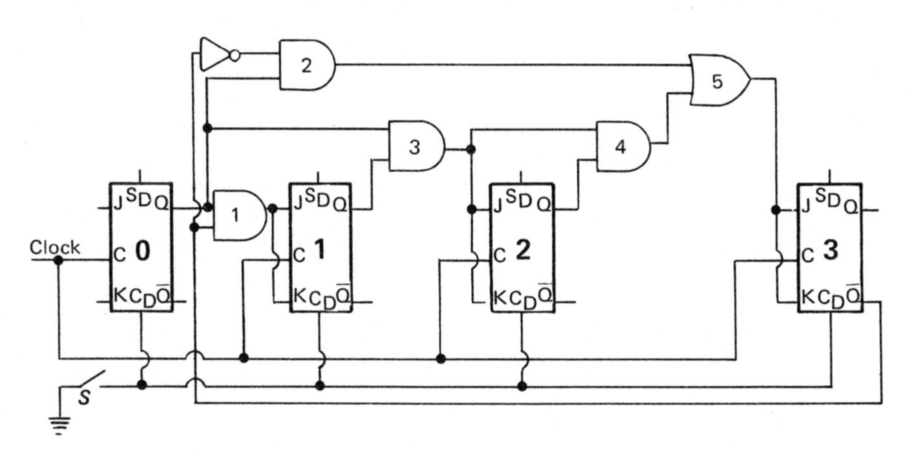

Fig. 3-4

Circuit diagram for a decade counter.

By rearranging the extra gates and keeping in mind the state of the flip-flops on the normal binary count, this counter could be made to count up to any number from 1 to 15. Any count that is a power of 2 requires no special gating.

The counters discussed here are up-counters; that is, they count from zero *up* to a specific number and repeat. Down-counters can be constructed similarly, except that the $\bar{Q}$ outputs indicate the count and are used to propagate the counting. Counters which count either up or down (called up/down counters) can also be constructed.

MONOSTABLE MULTIVIBRATOR

Very often it is desirable to generate a pulse of known width and shape in response to signals which may be of variable widths and of nonuniform shapes. A device extensively used for this purpose is a monostable multivibrator or "one-shot" (Fig. 3-5) which gives a single pulse (one-shot) of a reproducible nature. The length of the pulse and its magnitude can be adjusted to whatever desired (within limits). Thus, its two major uses are to provide uniform pulses and known delays.

This monostable multivibrator or "one-shot" operates quite simply. When the normally grounded switch S is opened, the output of the AND gate becomes 1, and the output of NOR gate B changes to "0." This brings the P input to NOR gate A to "0," causing the Q output to change to a logical 1. Voltage from the +5 V connected to NOR gate A through the resistance begins to charge the capacitor back up to +5 V, at which point Q returns to 0.

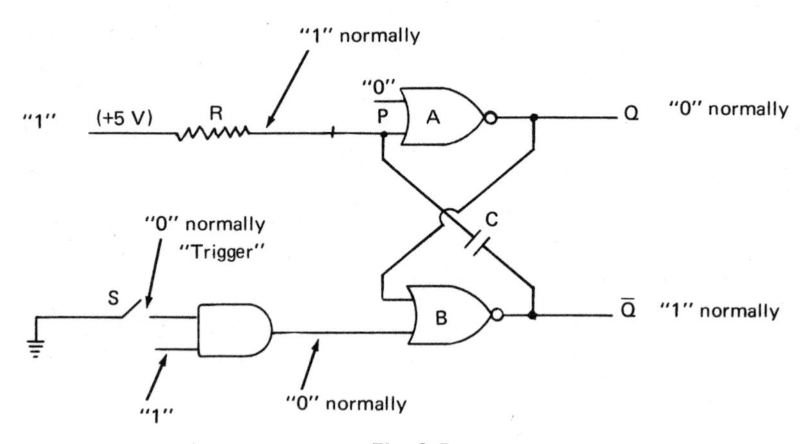

Fig. 3-5
Monostable multivibrator (one-shot).

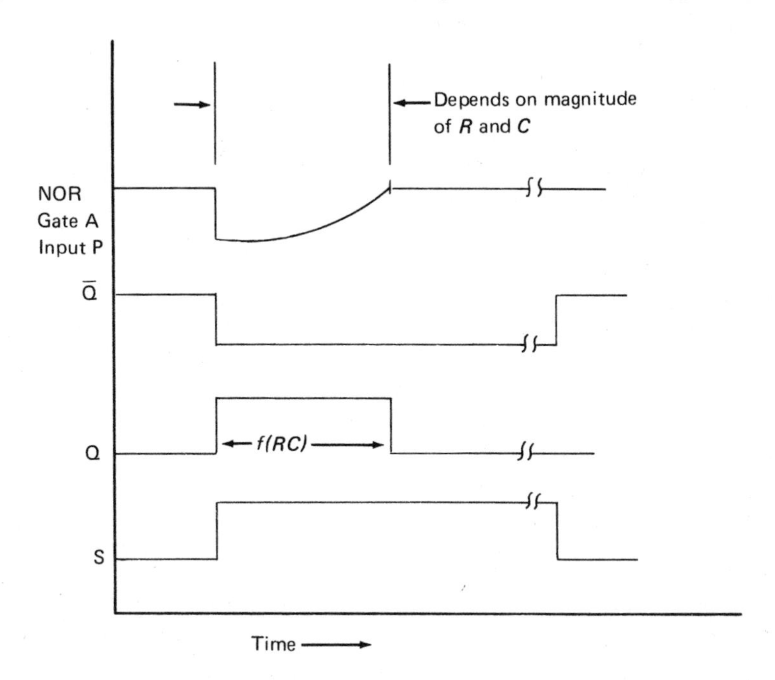

Fig. 3-6
Timing diagram for "one-shot" of Fig. 3-5.

Although $\overline{Q}$ does not return to zero until S is grounded again, the transition in Q is triggered only by a *negative-going transition* in $\overline{Q}$. Continued existence of a zero level $\overline{Q}$ will not cause additional pulses; hence the name "one-shot." The length of time required for this process is dependent on the values of the resistance and capacitance, or the RC time constant. (A value of 100 Ω and 0.01 μF will give a pulse duration of about 1 μsec.) The circuit could be changed to give a $1 \rightarrow 0 \rightarrow 1$ pulse, or an inverter could be used, on the Q output. The primary purpose of such a device is to give a single reproducible pulse or signal. Figure 3-6 shows the timing diagram for the one-shot of Fig. 3-5.

DATA LATCHES AND SHIFT REGISTERS

Two other devices which are much used in computer work are data latches and shift registers. Both of these devices can be constructed from simple clocked RS flip-flops. The data latch (also called a "buffer register") can provide a "buffered" connection between digital devices such as a computer and an external instrument. Its purpose is to temporarily store

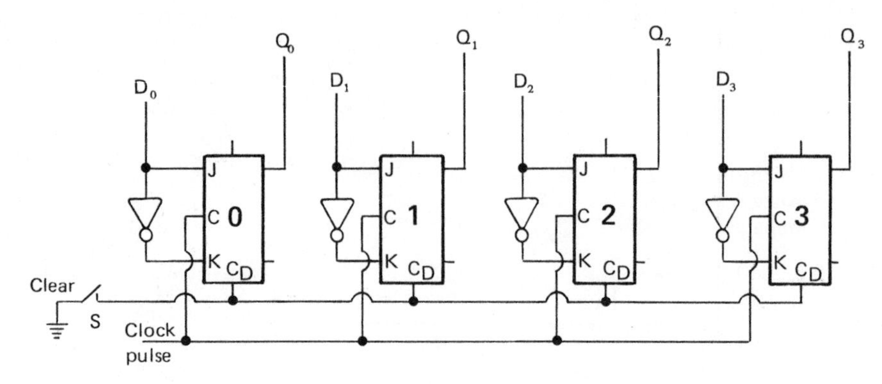

Fig. 3-7
4-bit data latch.

information in the form of a series of binary states. It can be constructed from any of the flip-flops (RS, clocked RS, or JK). Figure 3-7 shows the diagram of a 4-bit data latch using clocked JK flip-flops. The data is entered on the D lines and read out at the Q outputs. The current status of the input lines will be held at the outputs when the clock changes from $1 \rightarrow 0$. The inputs can only be entered in parallel; that is, all data lines are "strobed" into the data latch on a single clock pulse. It is desirable to use master–slave flip-flops so that the outputs can change state only at the instant that the $1 \rightarrow 0$ clock transition occurs.

The shift register is used also for temporary data storage. However, it is more versatile. It is, as the name implies, a register in which a bit pattern can be entered and shifted from flip-flop to flip-flop. The data can be entered either in serial or in parallel, shifted, and output in either serial or parallel, depending on the particular register. The desired method of data entry and output causes the circuit to be different from the data latch. It is mandatory

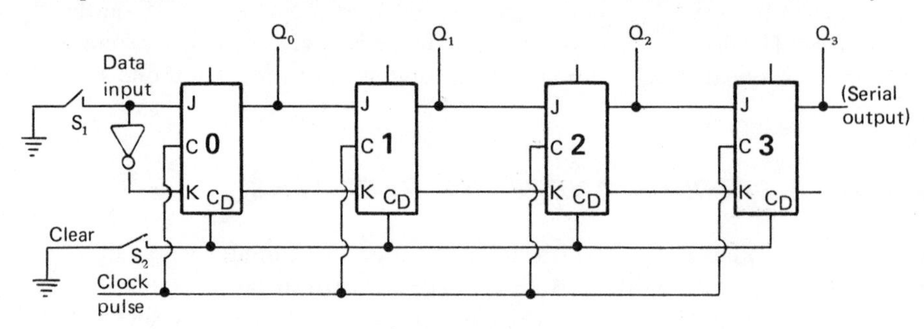

Fig. 3-8
4-bit shift register.

that master–slave flip-flops be used in the construction of a shift register. If any other is used, a bit entered at FF1 would ripple down the shift register while the clock state was true. This is prevented by using the master–slave configuration for each flip-flop. Figure 3-8 shows a simple serial input, parallel or serial output, 4-bit register. Here, the data is input one bit at a time, and on each clock pulse is shifted to the next flip-flop. After four clock pulses, the first bit entered in FF0 would be at FF3. Successive clock pulses will then move additional data bits through, outputting serially at Q_3. For n bits to be entered and shifted through requires $n + 4$ clock pulses. An alternative is to enter four bits serially and take them out in parallel at $Q_0, \ldots, Q_3$ after the fourth clock pulse. The number of bits which can be output in parallel is, of course, limited to the number of flip-flops in the register. Commercially, 4- and 8-bit (4 and 8 flip-flop) shift registers are available in integrated circuit packages.

EXPERIMENTAL

This experiment is designed to acquaint you with the building of various counters, buffer registers and data latches, shift registers, and monostable multivibrators. You should construct at least one type of each of the devices and observe their operation.

Equipment Needed

- 1 EL Digi-designer
- 2 Dual-in-line JK flip-flop packages (SN 7476)—(4 for option F4)
- 1 Quad 2-input NOR gate package (SN 7402)
- 2 Quad 2-input AND gate packages (SN 7408)
- 1 Hex inverter package (SN 7404)
- 1 Quad 2-input OR gate package (Signetics SP 384A)
- 2 NAND gates for each switch you wish to buffer (usually one 7400 package is sufficient)
- 1 Dual-trace regular oscilloscope, OR
- 1 Dual-trace storage oscilloscope (should have scan rate of at least 100 nsec/div.)
- 1 Variable resistance substitution box (15 Ω → 10 MΩ)
- 1 Variable capacitance substitution box (0.0001 → 0.22 μF)

A. Ripple Carry or Asynchronous Counter

1. Construct the circuit shown in Fig. 3-1. Be sure to consult the data sheet accompanying the JK flip-flop package. Wire a light to each

of the Q outputs. In order to compare the results with Table 3-1, it is suggested that the outputs be wired in the following positional order: Q_3 leftmost, Q_2 next, Q_1 next, and Q_0 rightmost.

2. Clear all the FF's (set Q's to 0) by momentarily grounding S. Pulse the C_0 input several times with a 0, 1, 0 pulse. After each pulse, observe the counter output. Compare the results with the entries of Table 3-1. What happens if an *un*buffered switch is used for pulsing?
3. Predict the state of the counter when 20 pulses have occurred. Provide 20 pulses, observe the counter, and compare the result with your prediction.
4. (Optional.) Connect channel 1 of the oscilloscope to output Q_0. Connect channel 2 to Q_3. Set FF's 0, 1, and 2 all at 1 (count of 7). Adjust the scope for internal trigger and a sweep rate of about 0.1 μsec. Now pulse C_0. Monitor the level changes for channels 1 and 2. What is the time delay? *Note*: You cannot see any delay if your oscilloscope is not able to scan at 100 nsec/div or faster.

B. Synchronous Counters

1. Construct the circuit shown in Fig. 3-3 Use lights to monitor the Q outputs. The lights should be wired as in Sec. A1. Also connect a light to each of the J inputs of FF2 and FF3.
2. Clear all FF's by momentarily closing S to ground.
3. Repeat A3.
4. (Optional.) Repeat A4. What was the time delay? How did it compare with A4 of the ripple carry or asynchronous counter?

C. Count-by-n

1. Wire the circuit shown in Fig. 3-4. Verify that it counts to 9 and returns to zero on the 10th pulse.
2. Using the appropriate gates, rewire the circuit to count to 12 (i.e., clear on the 12th pulse). It is suggested that you refer to the background discussion on synchronous counters for hints.
3. Prepare a diagram of a circuit for a synchronous counter which would count by 5, and one which would count by 8.

D. Monostable Multivibrator (One-Shot)

1. Construct the circuit shown in Fig. 3-5. Connect the normally 0 input of the AND gate through a toggle switch to ground. Use the 100 Ω and 0.01 μF settings of the substitution boxes. Connect the Q output to channel 1, and point P to channel 2 of the oscilloscope used. Use a sweep rate of 0.2 msec/div.

2. If a storage oscilloscope is available, open switch S and observe the effects on the scope. Repeat this procedure several times. If a storage oscilloscope is *not* available, connect the one-shot trigger input to a square-wave generator running at 1 KHz or faster. (The Digi-designer clock works well for this; alternately, the Schmitt trigger from experiment 1 could be used.) When the oscilloscope sweep trigger input is connected to the square-wave generator, it is possible to synchronize the observation of the voltages at Q and point P. How long does the Q output remain at 5 V? What does the signal from point P look like?
3. Change the resistance and capacitance settings. Repeat step 2 for several different settings. Are there any limitations on the values of R and C?

E. Data Latch or Buffer Register

1. Construct the circuit shown in Fig. 3-7 using JK flip-flops. Connect each data input line through a toggle switch to ground. Monitor the D inputs and Q outputs with lights.
2. Set various combinations of bits on the D lines; "strobe" the data in with a clock pulse. Note that Q outputs remain unchanged until the clock pulse occurs, after which they will have the corresponding D values. Save this circuit for the next section.

F. Shift Registers

1. Construct the circuit shown in Fig. 3-8 using JK flip-flops. Monitor each output with a light. Connect the "data in" line through a toggle switch (S_1) to ground.
2. Clear all FF's. Leave the S_1 switch open. Provide up to 4 successive clock pulses. Note the state of each FF after each pulse.
3. Clear all FF's and alternate 1 and 0 by opening and closing S_1 after each clock pulse. Note the effect each clock pulse has. Shift a 4-bit pattern completely into and out of the shift register.
4. (Optional.) (1) Hook the Q outputs to the D lines of the data latch or buffer register of Sec. E. (2) Shift into the shift register a 4-bit pattern. Now "strobe" this pattern into the data latch with a clock pulse to the data latch. Compare the Q outputs of the shift register to the data latch. (3) Repeat with a different 4-bit pattern.

Experiment 4

Other Useful Solid-State Devices*

FIELD-EFFECT-TRANSISTOR SWITCHES

In handling analog data signals, there are several important considerations. One problem is how to switch a data signal on and off at the speed and reliability level required by a computer. A second is how to compare voltages; and a third is how to amplify and/or sample analog signals for computer data acquisition.

Consider the problem of switching a signal on and off at a high rate. What are the characteristics of an ideal switch? What kinds of components most adequately meet these characteristics? First, the switch should have infinite resistance when off and zero resistance when on. Second, it should instantaneously and reproducibly open or close. Third, it should be easily controlled by electronic signals. Fourth, it should be stable and capable of switching both high and low voltages and currents with no change in signal level. Fifth, the ideal switch should be noise-free.

An ordinary toggle switch fulfills the first, second, and fourth criteria, but badly fails the third and fifth. A relay fulfills all these requirements to some degree, but is not capable of switching at high rates. In addition, relays have limited life spans. Diode and transistor switches could be used but their on/off resistances are not ideal.

The field-effect-transistor (FET) switch approaches the ideal requirements. It is a solid-state device which can be turned on or off at a very high rate and is quite reliable. It is turned on or off depending upon the voltage level applied at a "gate" terminal. Figure 4-1 shows the basic configuration of an FET switch. As long as the gate is set at a zero or positive voltage, the switch is "closed" (i.e., a low impedance is seen from source to drain), and a signal can traverse the switch unimpeded. If a sufficiently negative voltage is

* Field-effect-transistor switches (FETS), voltage comparators, and track-and-hold amplifiers.

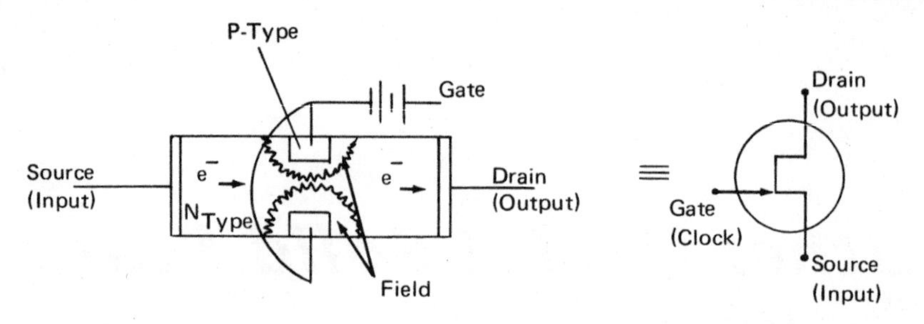

Fig. 4-1
Field-effect-transistor (FET) switch. When a negative voltage is applied to the gate, the field created inhibits electron flow, and the FET is off. When a positive (or zero) voltage is applied to the gate, the FET is on, allowing electrons to flow.

applied to the gate, then the field set up within the gate inhibits the transmission of the signal (i.e., shows a very high impedance), and the switch is "open." The gate can then be controlled by a (+5 to −5 V) clock input such that the switch is on when the clock is at +5 V and off when it is at −5 V. The FET has two disadvantages. Only a normally-on operation is available, and there is significant gate current. The metal–oxide–semiconductor FET (MOSFET) overcomes these problems but is less stable than the FET. The FET is, however, the most widely used, as it best fulfills the requirements needed in computer-controlled switching. The FET recommended in this experiment is actually a monolithic integrated circuit, which is *internally biased*, so that it can operate from normal TTL logic levels (i.e., off at *zero* volts, and on at +5 V).

COMPARATOR

A second important device is a voltage comparator. Its basic operation is to output one of two states, a logical 1 or 0, depending upon the magnitude of an unknown voltage compared with a known or reference voltage. A

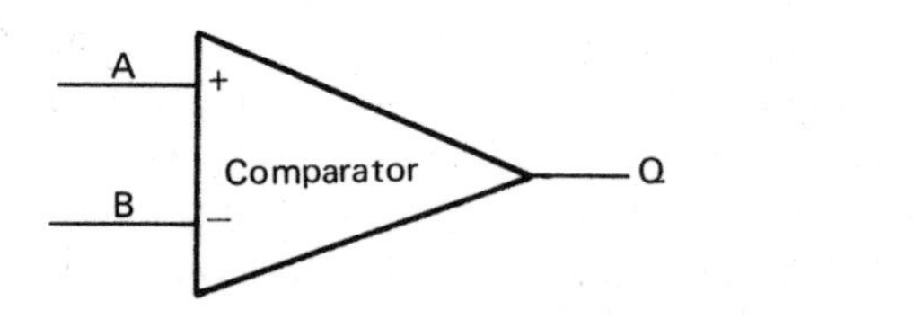

Truth Table

A,B	Q
A > B	1
A < B	0

Fig. 4-2
Symbol and truth table for comparator.

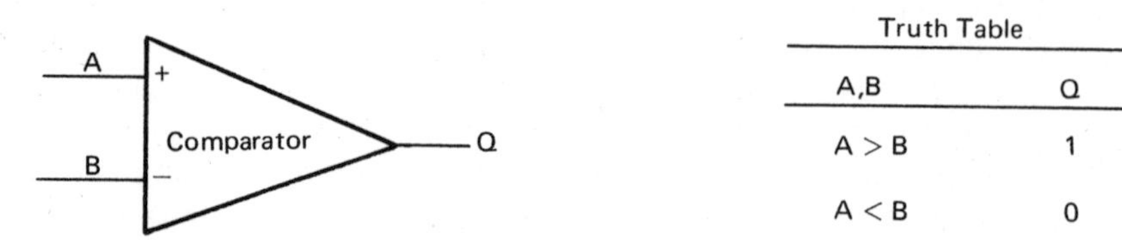

comparator is, therefore, a difference amplifier with a binary output. Figure 4-2 shows both the typical symbol and the truth table for one comparator. These devices will switch states any time the ΔV of A, B is greater than about 1 mV. They can typically handle input ranges of 0 to ± 10 V. The comparator output can be used to control other events depending on whether the input voltage is less than or greater than the reference voltage.

TRACK-AND-HOLD AMPLIFIERS

Track-and-hold (T/H) amplifiers are solid-state amplifiers which have the capability of following a signal and subsequently holding a particular value for a period of time. The track-and-hold amplifier is a simple solid-state amplifier used in a voltage follower mode, i.e., the output is tied back to the inverting input, the signal is fed to the noninverting input, and a capacitor of suitable size is connected between the signal and ground. (For a brief review of operational amplifiers, see Appendix H.) Figure 4-3 shows a diagram of a track-and-hold amplifier. It operates in the following manner: As long as signal is coming from the signal line, the capacitor charges. If the signal is switched off, the capacitor ceases to charge and begins to slowly discharge. The charge drives the operational amplifier; if the capacitor is of proper size, the output of the operational amplifier mirrors the input signal at the moment the signal line is switched off.

Figure 4-4 shows a timing diagram of the input, output, and clock signals for a T/H amplifier. When the clock input to the FET is low, the input signal

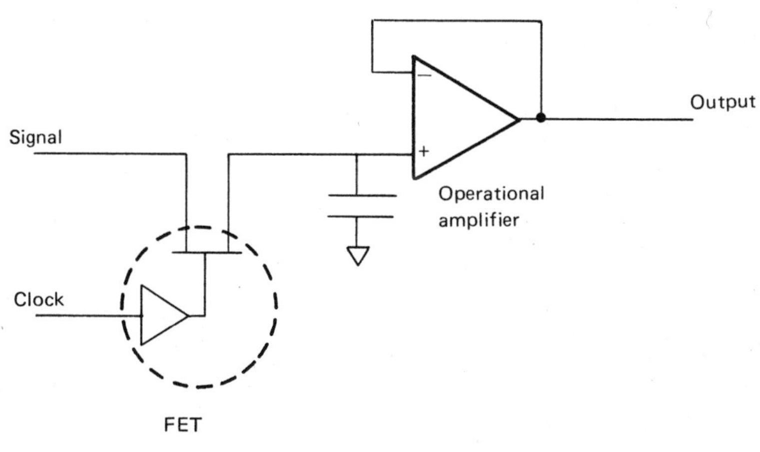

Fig. 4-3
Simple track-and-hold (T/H) amplifier.

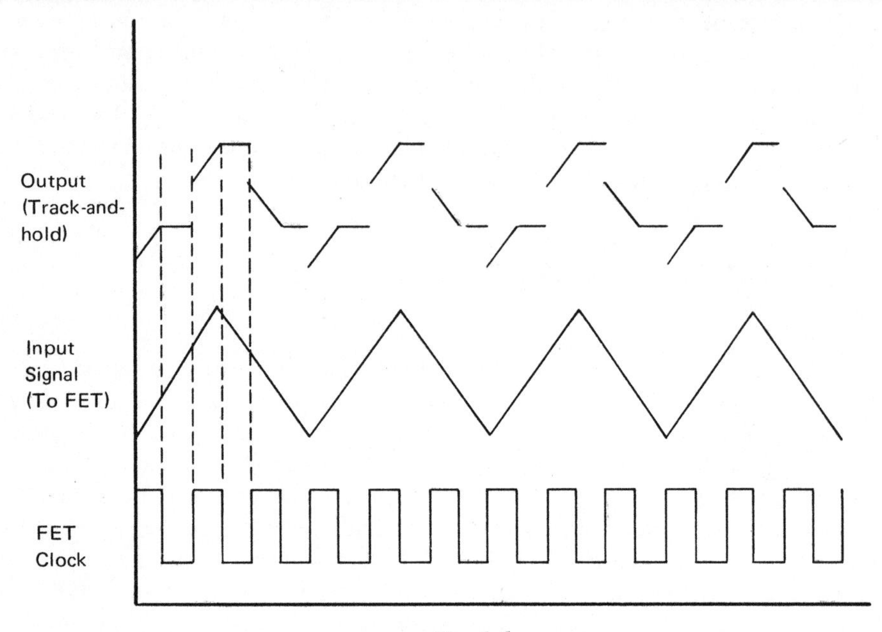

Fig. 4-4

Theoretical track-and-hold signals.

is turned off, and the output of the T/H "holds." When the clock is high, the FET is turned on, and the amplifier again follows (or "tracks") the input signal. The rate of switching the amplifier to the "hold" mode should be such that the actual signal will change very little during this time.

The rate at which the capacitor loses its charge, giving rise to error in the output signal of the operational amplifier, is called the "droop rate." This droop rate depends upon the magnitude of the capacitor used. T/H amplifiers can be purchased commercially in IC packages and have much better characteristics than shown here. As a result, you would not ordinarily wire the circuit shown from individual components.

In this experiment, you will observe the operation of a FET switch, voltage comparator, and track-and-hold amplifiers in several configurations.

EXPERIMENTAL

Equipment Needed

1 EL Digi-designer
1 ±15-V power supply

4 Voltage reference sources (variable 0 to 15 V)
1 Field-effect-transistor switch (Teledyne Crystalonics CAG-30. If a substitute is used, make sure that the gate operates on normal TTL logic levels.)
1 Voltage comparator (LM 710C)
2 BCD Counters (SN 7490)
1 Operational amplifier (SN 72741)
1 Function generator (see Appendix E)
1 Pulse generator (variable rate)
1 Dual-trace storage oscilloscope
1 Volt–ohm multimeter (VOM)
1 Toggle switch
5 Capacitors: one each of 5, 50, and 500 pF, and two of 0.01 μF

A. Field-Effect-Transistor Switch (FET)

1. Consult the data sheet for the FET switch (CAG-30) and power it. Connect the pulse generator (set at $\simeq$1 KHz) to the signal input pin. Connect the gate (clock) input pin to +5 V through a buffered toggle switch. Connect the output of the FET switch to channel 1 of the oscilloscope. Connect channel 2 to the buffered switch output. The oscilloscope trigger should be connected to the pulse generator.
2. Alternately open and close the switch which applies +5 V to the gate. Monitor and record the output of the switch. If a storage oscilloscope is not available, connect the gate pin to a square-wave source and view the repetitively triggered scan.
3. Remove the +5-V source from the gate pin. Connect a variable voltage reference source to it. Set the source at 1, 2, 3, and 4 V, each time repeating step 2.

B. Voltage Comparator

1. Consult the data sheet for the LM 710C voltage comparator. Notice particularly the maximum acceptable input voltages. Take care not to exceed them, or you may damage the comparator. Power it according to this data sheet, using 0.01 μF decoupling capacitors between V_+, V_- and ground. Set one VRS to −4 V and connect to the negative terminal of the comparator. Hook a second VRS to the + side of the 710 comparator. The second VRS should be set at −0.5 V. Monitor the output of the VRS with the VOM. Monitor the output of the comparator with a light.
2. Vary the voltage on the positive side of the comparator (−0.5, −1.0, −1.5, −2.0, −2.5, −3.0, −3.5, −4.0), noting the effect on the

output of the comparator. At what point does a change occur in the output?

3. Use smaller steps in the input voltage and note the sensitivity of the change of the voltage comparator. Would the addition of a flip-flop help to stabilize the output signal?

C. *Track-and-Hold Amplifiers*

1. Wire the two SN 7490 counters as shown below.

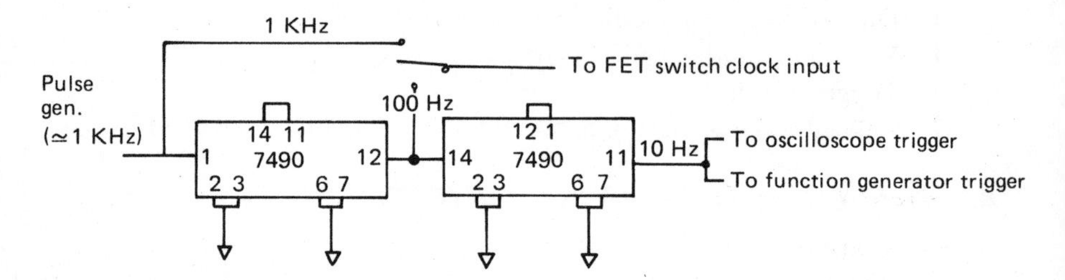

Connect the 10-Hz output of the counter to the function-generator and oscilloscope triggers. Wire the circuit in Fig. 4-3, using a 0.01 μF capacitor. Connect the triangular wave output of the function generator to the signal-input pin of the FET switch, and connect the toggle switch (100 Hz or 1 KHz) to the FET clock input. Check the data sheets to be sure that all chips are properly powered.

2. Set the dual-trace scope at 2 V/cm vertical and a sweep rate of about 10 msec/div. Set the toggle switch to the 100 Hz position and monitor the input and output signals of the track-and-hold amplifier with the scope. Does the output of the amplifier hold the signal? Does the input signal change very much while the amplifier is in the hold mode?
3. Set the toggle switch at the 1 KHz position. Repeat step 2.
4. Set the scope on the storage mode. Remove the capacitor between the FET switch and the amplifier. Set the toggle switch to 100 Hz. What happens to the output signal when the amplifier is in the hold mode? Now insert, successively, 5-, 50-, 500-pF capacitors and finally the 0.01-μF capacitor. What happens as the capacitors become larger? Does the "droop" rate become very slow as the capacitors get larger?

Experiment 5

Properties and Use of Digital-to-Analog (DAC) and Analog-to-Digital (ADC) Converters

DIGITAL-TO-ANALOG CONVERSION

Most laboratory instruments either produce or accept analog or continuous signals of some form. In order for a digital computer to be able to communicate with an analog instrument, digital–analog interconversions must be made. There are several different ways to construct devices to accomplish this. A digital-to-analog converter (DAC) can be constructed using a weighted resistor or ladder resistor network. The ladder resistor network is the more frequently used because it requires only two different values of precision resistors and is therefore cheaper to construct. However, the weighted resistor network is easier to conceptualize, and will be discussed first. Figure 5-1 shows the diagram of a typical four-bit DAC, constructed from a weighted resistor network. At any particular time, the value of the current at the inverting input of the operational amplifier is inversely proportional to the values of the resistors in the circuit and directly proportional to V_R. The (inverted) output voltage, V_{out}, depends on these parameters and on the value of R_f. The exact relationship is shown in Eq. (5-1), where L_i denotes the logical state of the switch adjacent to the ith resistor. The quantity L is 1 for closed switches and 0 for open switches.

$$V_{out} = -iR_f = -\left[V_R\left(\frac{L_3}{R_3} + \frac{L_2}{R_2} + \frac{L_1}{R_1} + \frac{L_0}{R_0}\right)\right]R_f \qquad (5\text{-}1)$$

If each succeeding resistor has a value one-half that of the preceding one, then it will conduct twice as much current. In this case, each resistor and switch combination represents a binary digit (bit) of digital-to-analog

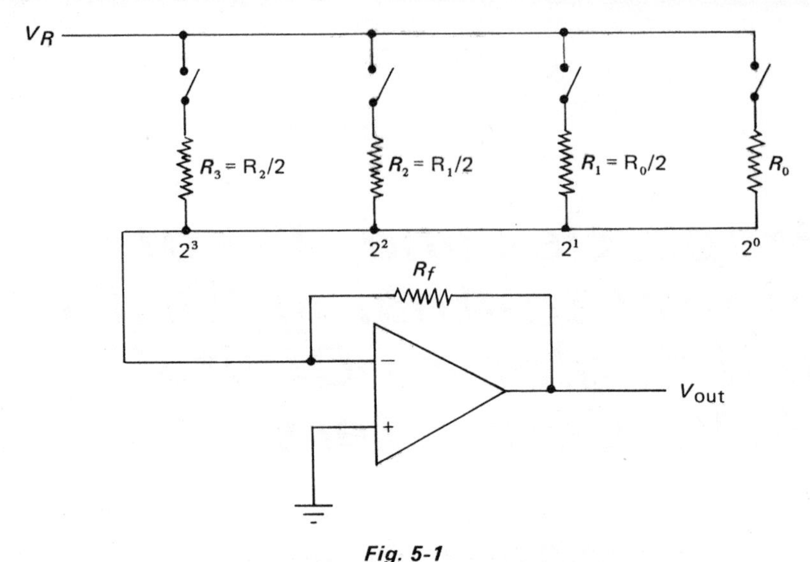

Fig. 5-1
A 4-bit DAC using a weighted resistor network.

conversion, where the least significant bit (2^0) is represented by the largest resistor, and the most significant bit (in this case, 2^3) by the smallest resistor. The total digital-to-analog conversion is then the summation of each closed switch–resistor combination in the circuit. Resistors with open switches make no contribution to the DAC. The resultant output voltage is then calculated from Eq. (5-1).

An example will clarify this. Suppose the reference voltage V_R is 15 V, R_0 is 100 kΩ, R_f is 6.66 kΩ, and binary number 1101 is to be converted with the 4-bit DAC. What will the output voltage V_{out} be? Equations (5-2a) to (5-2c) show the step-by-step evaluation of Eq. (5-1), to yield a value of −13 V for V_{out}.

$$V_{out} = -\left[V_R\left(\frac{1}{R_3} + \frac{1}{R_2} + \frac{0}{R_1} + \frac{1}{R_0}\right)\right]R_f \tag{5-2a}$$

$$V_{out} = -\left[15\left(\frac{1}{12.5\ \text{k}\Omega} + \frac{1}{25\ \text{k}\Omega} + \frac{0}{50\ \text{k}\Omega} + \frac{1}{100\ \text{k}\Omega}\right)\right]6.66\ \text{k}\Omega\ \text{V} \tag{5-2b}$$

$$V_{out} = -[100(0.08 + 0.04 + 0.01)] = -13\ \text{V} \tag{5-2c}$$

Table 5-1 shows the output values for all possible combinations of switch closures, for the component values given above. The DAC can be constructed to utilize a larger number of bits by continuing the decrease in magnitude of the resistors, using as many resistors as you desire bits. It should be pointed

Table 5-1

Table of Values ($\times 10^3$) of all Components for Fig. 5-1 when $V_R = 15$ V, $R_0 = 100$ kΩ, and $R_f = 6.66$ kΩ

Decimal	Binary $2^3 2^2 2^1 2^0$	L_3/R_3	L_2/R_2	L_1/R_1	L_0/R_0	$1/R_{tot}$	I, mA (V_R/R_{tot})	$-E_{out}$, V ($I \cdot R_f$)
0	0000	0	0	0	0	0	0	0
1	0001	0	0	0	0.0100	0.0100	0.150	−1.00
2	0010	0	0	0.0200	0	0.0200	0.300	−2.00
3	0011	0	0	0.0200	0.0100	0.0300	0.450	−3.00
4	0100	0	0.0400	0	0	0.0400	0.600	−4.00
5	0101	0	0.0400	0	0.0100	0.0500	0.750	−5.00
6	0110	0	0.0400	0.0200	0	0.0600	0.900	−6.00
7	0111	0	0.0400	0.0200	0.0100	0.0700	1.050	−7.00
8	1000	0.0800	0	0	0	0.0800	1.200	−8.00
9	1001	0.0800	0	0	0.0100	0.0900	1.350	−9.00
10	1010	0.0800	0	0.0200	0	0.1000	1.500	−10.00
11	1011	0.0800	0	0.0200	0.0100	0.1100	1.650	−11.00
12	1100	0.0800	0.0400	0	0	0.1200	1.800	−12.00
13	1101	0.0800	0.0400	0	0.0100	0.1300	1.950	−13.00
14	1110	0.0800	0.0400	0.0200	0	0.1400	2.100	−14.00
15	1111	0.0800	0.0400	0.0200	0.0100	0.1500	2.250	−15.00

out that the voltage reference source V_R must have a stability of $\pm\frac{1}{2}$ of the voltage output of the least significant bit; i.e., for the DAC shown in Fig. 5-1 with the values given in Table 5-1, the stability must be at least ± 0.5 V.

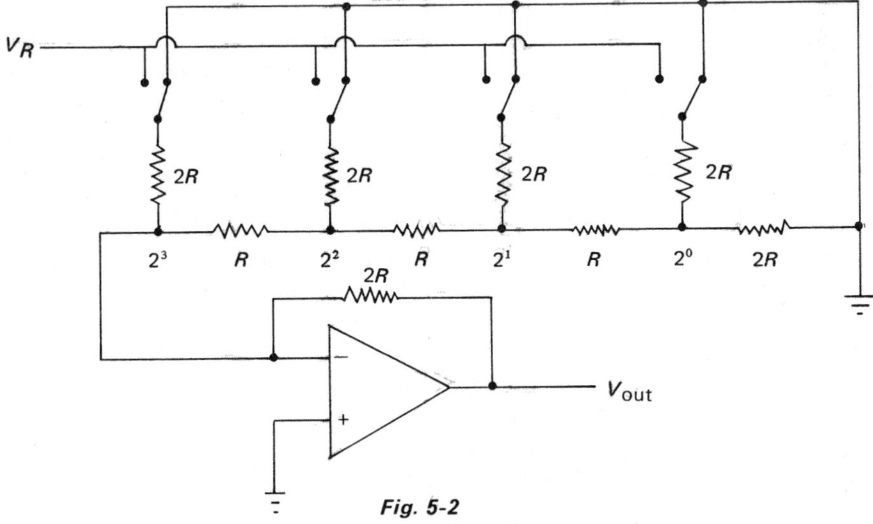

Fig. 5-2
A 4-bit DAC using a ladder resistor network.

A second type of DAC is one constructed from a ladder resistor network. Figure 5-2 shows such a 4-bit DAC. The switch "closest" to the amplifier (the leftmost one in Fig. 5-2) corresponds to the most significant bit, while the one "farthest down the ladder" corresponds to the least significant bit (the rightmost in this case). Although the resistance ladder is less straightforward than the weighted resistor network, Kirchhoff's laws can be used to solve for the output voltage, yielding Eq. (5-3)*:

$$V_{out} = -iR_f = -\left[V_R\left(\frac{L_3}{2R} + \frac{L_2}{4R} + \frac{L_1}{8R} + \frac{L_0}{16R}\right)\right]2R$$

$$= -V_R\left[L_3 + \frac{L_2}{2} + \frac{L_1}{4} + \frac{L_0}{8}\right] \tag{5-3}$$

Here, L_i is 0 if the ith switch is *grounded*, and 1 if it is connected to V_R. Thus, if V_R is 4 V, the binary number 1101 will generate a −6.5 V output. Additional bits can be constructed by adding ladder sections. The power-supply stability requirements for a ladder DAC are the same as for a weighted resistor network, one-half the voltage output of the least significant bit.

ANALOG-TO-DIGITAL CONVERSION

There are a great many types of commercially available analog-to-digital converters (ADC's). A few of these are the staircase–ramp converter, the continuous-balance converter, the successive-approximation converter, the continuous-ramp converter, and the dual-slope integrating converter. All of these operate by comparing the unknown voltage (V_u) against a known reference voltage V_R; they differ principally in the mechanism used for generating V_R and in the time required for a complete conversion.

The first of these, the staircase–ramp voltage-comparison converter, is shown in block diagram form in Fig. 5-3a. This analog-to-digital converter operates in the following way: When an ADC reset pulse occurs, the Q output of the RS flip-flop is set to 1, and the counter starts to count up as it receives clock pulses. The digital counter output is input to a DAC which generates a reference voltage V_R. This reference voltage is then compared against V_u in the voltage comparator. When the generated reference voltage becomes slightly larger than V_u, the RS flip-flop is reset, disabling the clock through the NAND gate. At this point, the counter contains a binary value proportional to the input voltage V_u, and the digitization step is complete. The counter contents can be transferred to a computer or data buffer and then cleared so

*For a derivation, see ref. A-13, pp. 334–335.

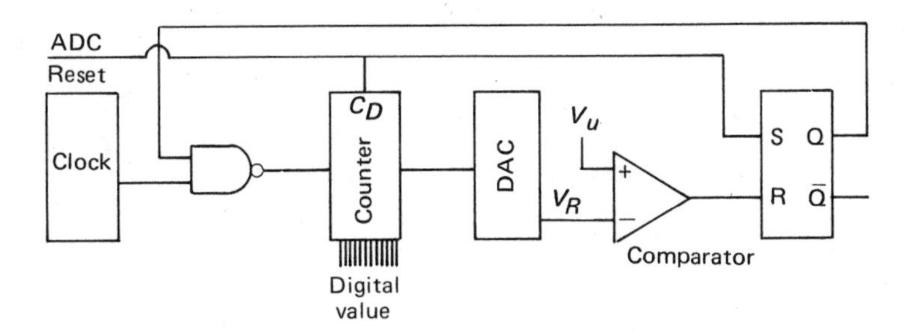

Fig. 5-3a

Staircase–ramp voltage-comparison ADC.

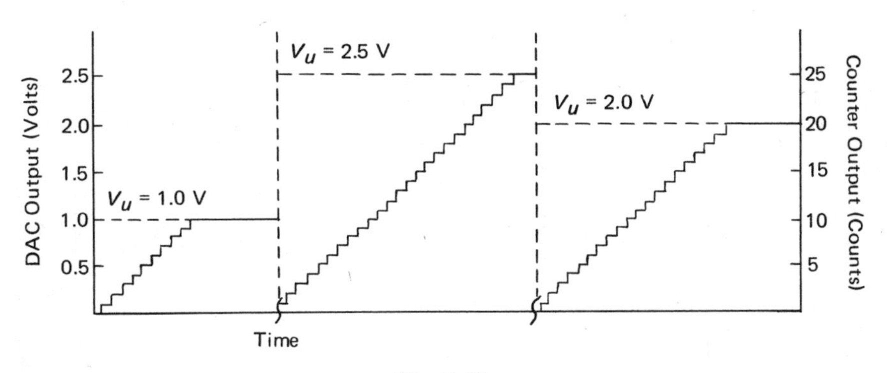

Fig. 5-3b

Comparison of DAC and counter output to V_u analog input with time for ADC of Fig. 5-3a.

the ADC is ready for another conversion. Figure 5-3b shows the relationship of the counter value with the unknown voltage for this ADC. In this case, one count is equal to 0.1 V, but this relationship will vary for other ADC's, depending on the number of bits and the dynamic range. Note that the conversion time depends on the magnitude of the unknown voltage.

The continuous-balance voltage-comparison converter block diagram in Fig. 5-4a represents a modification of the staircase–ramp ADC. The major difference is that a counter which can count *either up or down* is used to control the DAC which provides reference voltage V_R. Whether the counter counts up or down depends on whether the comparator finds V_u larger or smaller than V_R. There is, of course, more gating circuitry to control the up/down counter. Also, there is often a circuit which provides a "conversion complete" signal every time the comparator output changes logic levels (not shown in Fig. 5-4a). Whenever the data latch clock is pulsed, the contents of the up/down counter are transferred to the latch. Between latch clock pulses, the

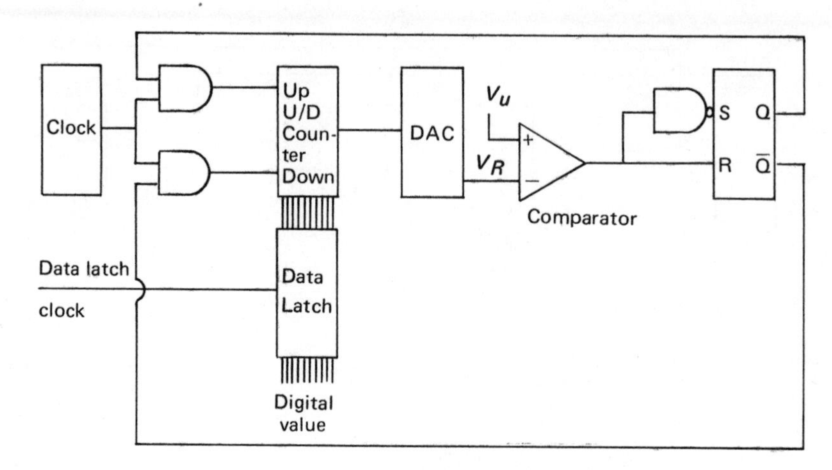

Fig. 5-4a
Continuous-balance voltage-comparison ADC.

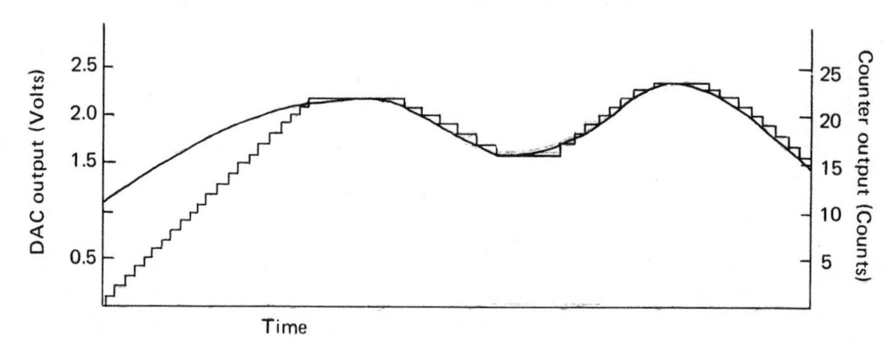

Fig. 5-4b
Comparison of DAC and counter output to analog input with time for ADC of Fig. 5-4a.

contents of the latch are constant and can be transferred to either a computer or some other device. Since the data latch can be clocked independently of the ADC clock, it is possible to externally synchronize both the A–D conversion, and the transfer of the digitized result. (Although the conversion process itself is actually continuous, the "conversion complete" signals show the time of the conversion and can be used to synchronize data acquisition.) Figure 5-4b shows a typical analog signal and the digital output for this type of ADC as a function of time. Although the conversion times are usually shorter than for the staircase–ramp ADC, they are still dependent on both the magnitude and the rate of change of the analog signal.

A block diagram of a successive-approximation ADC is shown in Fig. 5-5a. This converter operates in the following manner: The voltage output of the DAC (V_R) is first compared to the unknown voltage when the most significant bit of the DAC is set to 1. If the V_u is greater than V_R, then that bit is left on and the next most significant bit is turned on and the voltage output is again compared. If V_u is less than V_R, this bit is turned off. *Successively*, each bit is turned on, the DAC output voltage is compared to the V_u, and the bit is left on if V_u is greater than V_R. This process continues until the least significant bit (LSB) is tested; at this time, the value of the register is the binary equivalent of V_u and can be read by the computer. This type of ADC is one of the fastest and most commonly used. It requires only about 10 to 50 μsec per conversion as opposed to several hundred microseconds for some other types. In addition, the conversion time is independent of the magnitude of V_u. This is shown in Fig. 5-5b, which depicts the relationship between the analog (input) and digital (output) signals for a 6-bit successive-approximation ADC.

The staircase–ramp, continuous-balance, and successive-approximation converter resemble one another in that all of them utilize a DAC which is

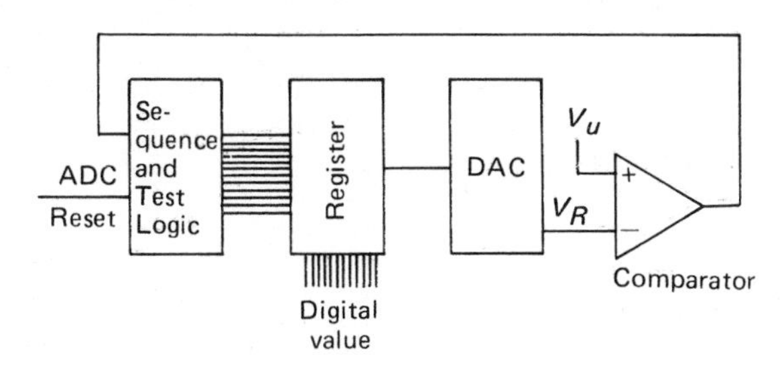

Fig. 5-5a
Successive-approximation voltage-comparison ADC.

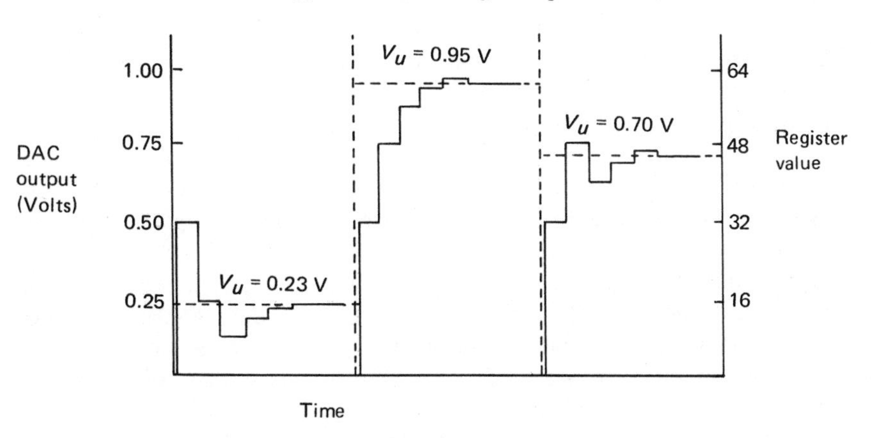

Fig. 5-5b
Comparison of DAC and counter output to analog input with time for ADC of Fig. 5-5a.

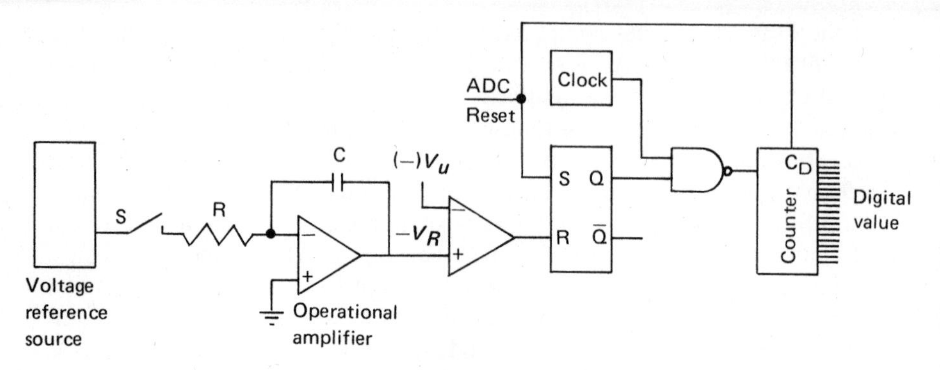

Fig. 5-6a
Continuous-ramp voltage-comparison ADC.

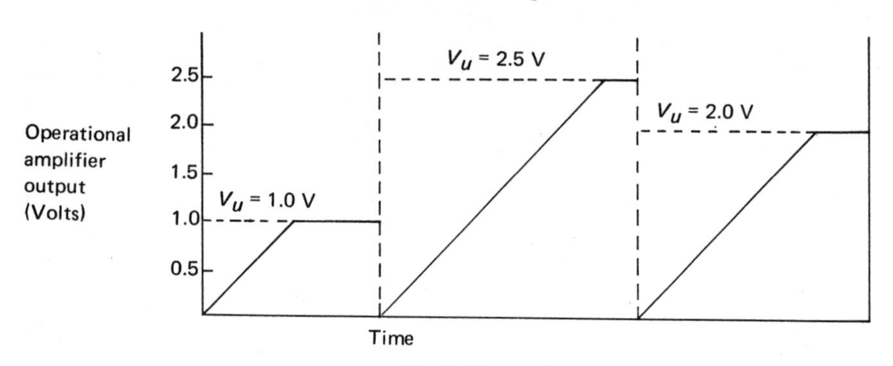

Fig. 5-6b
Comparison of V_R with V_u for ADC of Fig. 5-6a.

controlled either by a counter or register to generate the reference voltage. They differ principally in the method used to increment the counter-control register. The continuous-ramp and dual-slope integrating converters use a different method to generate V_R.

The continuous-ramp voltage comparison converter is shown in Figure 5-6a. When the ADC reset line is momentarily grounded, and switch S is closed, the clock is enabled and the counter begins to count up. Initially, the comparator will be in a logical one state. The output of the operational amplifier increases linearly to the limit of the voltage reference source. When the output of the amplifier slightly exceeds that of the analog input to the comparator, V_u, the comparator output will change from a 1 to a 0 state, causing the RS flip-flop to be reset. This, in turn, causes one of the inputs to the NAND gate to become a 0 and to stop the counter.

The counter values will depend on both the clock rate and the rate of increase of the ramp voltage, which depends on the RC value. The ADC must be carefully calibrated before use. Since the ramp voltage generated is linear, the counter output is directly proportional to the voltage of the ramp at any particular time. The values of R, C, and the clock rate are selected so that the counter simultaneously reaches its maximum when the ramp voltage reaches its maximum value.

Figure 5-6b shows a plot of the ramp voltage (only) versus time, for several analog voltages. Although the ramp voltage is continuous, the counter output, by its very nature, is *discontinuous*. If the *counter value* is plotted against time, a staircase function identical to that shown in Fig. 5-3b (counter scale only) is obtained.

The dual-slope integrating ADC operates on a principle quite different from those discussed above. A diagram of such an ADC is shown in Fig. 5-7. This converter operates by comparing two times: (1) the time it takes to charge a capacitor with the unknown voltage V_u (this time is fixed and is simply the time required for the counter to reach its maximum value); and (2) the time required to discharge the capacitor when it is attached to the known voltage reference of opposite polarity, V_R.

In operation, the sequence of events is straightforward. When the converter is started by pulsing "ADC Reset," the Q output of the JK flip-flop is zero, and V_u is connected through R to capacitor C, which begins to charge. Immediately, the output voltage of the operational amplifier becomes greater than zero, the comparator changes state, and the gate opens allowing clock pulses to reach the counter while the capacitor continues to charge. The counter, whose most significant bit (MSB) is connected to the clock input of the JK, runs through its maximum value and, when it resets to zero, causes the JK to "toggle." This, in turn, causes V_R to be switched out and a known fixed reference voltage *of opposite polarity* to be switched into the circuit. The comparator has not changed state, so the counter is still operating. The capacitor now discharges at a fixed known rate through the comparator.

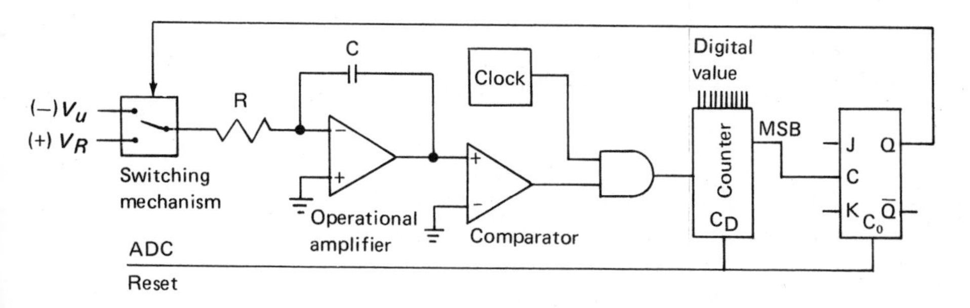

Fig. 5-7
Dual-slope integrating, ADC.

When the capacitor has fully discharged, the comparator changes state, disabling the gate and stopping the counter. The charge on the capacitor is directly related to the magnitude of V_u since the time of charging is always the same (i.e., the maximum count time of the counter). The discharge *rate* is fixed by V_R, and the time required for total discharge is proportional to V_u. Thus, if t_2 is the discharge time, t_1 is the charge time, and V_R equals reference voltage, then $V_u = (t_2/t_1)V_R$. For a given ADC, t_1 and V_R are known or can be adjusted to give t_2 in known units. The value of the counter can be stored in a data latch or read directly into the computer. Alternately, a visual display of the count allows a direct reading of a number proportional to the unknown voltage.

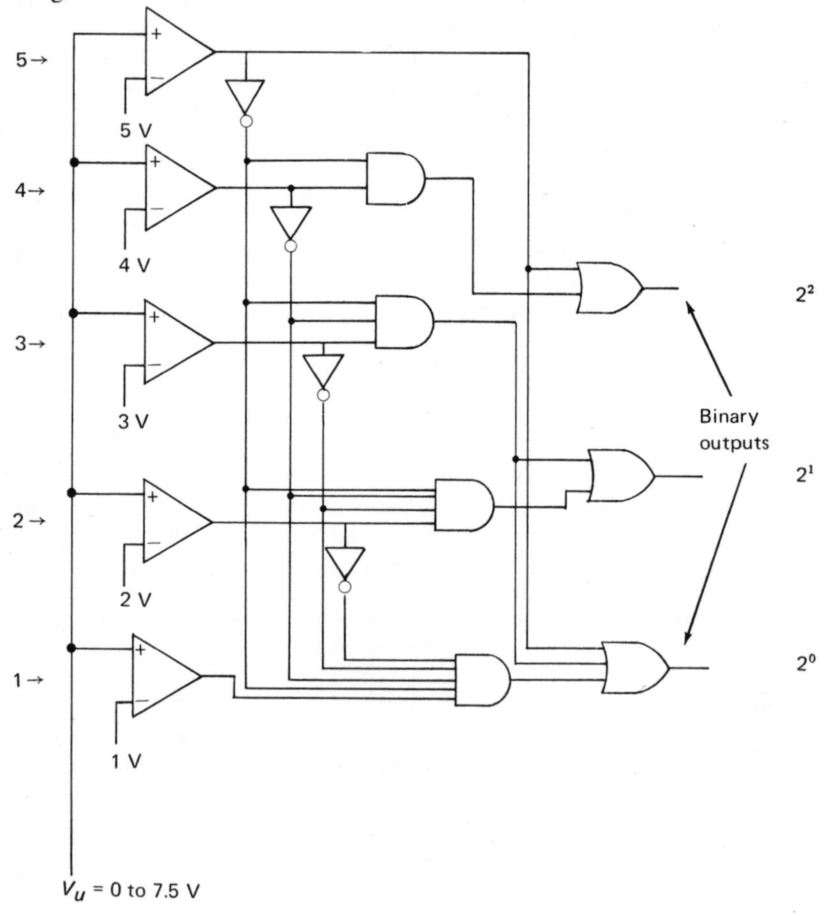

Fig. 5-8
A 3-bit ADC from gates and comparators.

It is possible to build a small 3-bit ADC using voltage comparators and gating. Such a circuit is shown in Fig. 5-8. It should be obvious, however, that this ADC becomes both expensive and unwieldy whenever the range of values to be converted is larger than 10:1.

The objective of the experiment is to acquaint you with several aspects of DAC's and ADC's. You will build and use several of the types described here.

EXPERIMENTAL

Equipment Needed

1 EL Digi-designer
1 ±15-V power supply
3 Voltage reference source (variable, 0–10 or 0–15 V)
1 Quad 2-input NAND gate package (SN 7400)
1 Hex inverter package (SN 7404)
1 Triple 3-input NAND gate package (SN 7410)
1 Bistable data latch (SN 7475)
1 BCD counter (SN 7493)
1 Up/down BCD counter (SN 74193)
1 Voltage comparator (LM 710C)
1 Operational amplifier (SN 72741)
1 Digital-to-analog converter (DATEL DAC-9)
1 Pulse generator or clock
1 Volt–ohm multimeter
9 1% resistors: 10 MΩ, 100 kΩ, 50 kΩ, 25 kΩ, 12.5 kΩ, 6.66 kΩ (a combination equal to this), two 2 kΩ, and 1 kΩ
13 Capacitors: one 1 μF and twelve 0.01 μF

Procedure

Be sure to read carefully the data sheets for each device. It is suggested that the IC pin layout be drawn on a separate sheet for the entire circuit.

A. Digital-to-Analog Converter (DAC)

1. Wire the diagram of Fig. 5-1 using 6.66 kΩ for R_f, 100 kΩ for R_0, and 15 V for V_R. Use the 72741 operational amplifier.
2. Connect the VOM to the output of the operational amplifier. Run the digital switches through all values from 0 to 15; monitor and record the output after setting each configuration of switches. Compare with the values in Table 5-1.

3. Repeat step 2 but monitor and record the current output at the operational amplifier input.
4. Draw a diagram for an 8-bit DAC, using the weighted resistor network technique.

B. *Analog-to-Digital Converter (ADC): Staircase-Ramp Voltage-Comparison Converter*

The complete circuit diagram for this ADC is shown in Fig. 5-9. You will put it together part by part, *testing it after addition of each part.* (Note that the comparator inputs are *opposite* those in the block diagram of Fig. 5-3a. This is necessary because the current-to-voltage conversion amplifier generates a *negative* V_R.) *Note*: If components other than the ones suggested are used, be sure to consult the data sheets for each, to be certain that they operate in a similar fashion.

1. Wire a 7493 counter using the data sheet provided. Attach the *A*, *B*, *C*, and *D* outputs to indicator lights. Connect a normally off pulser to the reset pin. Don't forget to connect the *A* output to the *B* input. Temporarily connect a normally off pulser to the *A* clock input pin. Check that the counter counts and resets properly.
2. Power a DAC-9 using pin 9 for $+15\ V$ dc from your power supply and pin 11 for ground. Check that ground and common are shorted

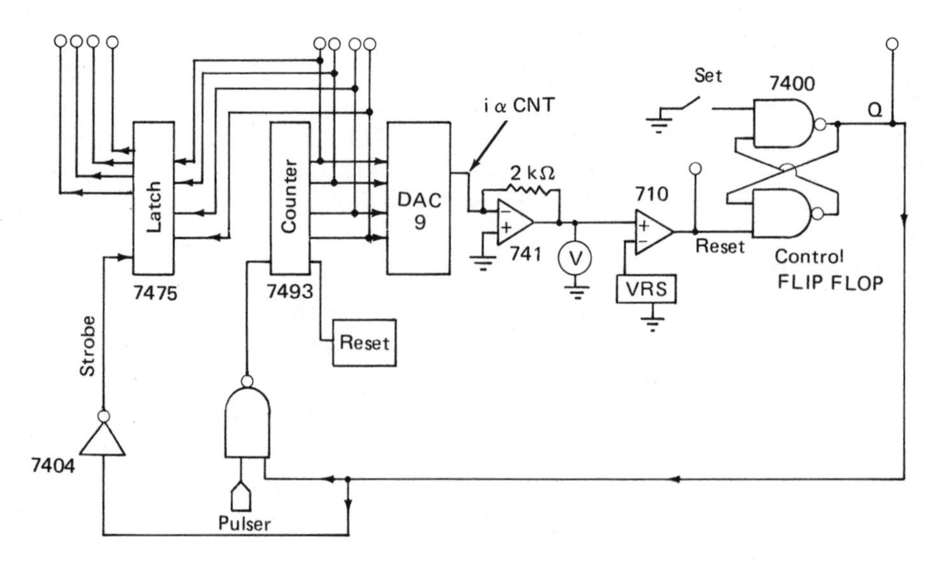

Fig. 5-9
Staircase-ramp voltage comparison converter.

on the power supply. Ground pins 5 through 8. (They are for the least significant bits and will not be used in this experiment.) Connect the *A*, *B*, *C*, and *D* counter outputs to pins 4, 3, 2, and 1 of the DAC. Pin 1 is for the most significant bit and pin 4 for the least significant bit. Take the output from pin 12. Connect it to your VOM using the 150 μA or 15 mA scale. With the counter at zero, your converter should produce no current; at a count of 15, the output should be about 2.4 mA. Make a plot of mA output vs count.

3. The current output of the DAC must be converted to a voltage. An operational amplifier is used to perform this function. (For a review of operational amplifiers, see Appendix H.) Wire a 72741 operational amplifier as shown in Fig. 5-10. In this mode, $V_{out} = -i_{in} \times R_f$. Since $R_f = 2\ \text{k}\Omega$, the maximum output is -5 V. Be sure to put decoupling capacitors across $+15$ V to ground and -15 V to ground for the operational amplifier. Now plot voltage output from the operational amplifier vs digital counts.
4. Because the comparator oscillates near the null point, another component must be included to generate a steady logical 0 at the comparator's transition point. Thus, the cross-coupled NAND gate (RS flip-flop) was used in Fig. 5-9. Set the flip-flop by grounding the set input *momentarily*, through a toggle switch. Connect the comparator to the reset input. Monitor the *Q* output of the flip-flop. When the comparator output changes from 1 to 0, this will change the state of the *Q* output from 1 to 0. Any oscillations at the reset input after the initial transition will not affect the *Q* output. Check the operation of this gate, using a normally off pulser in place of the comparator. Use decoupling capacitors for the 7400's between V_{cc} and GND.
5. Connect the RS flip-flop's *Q* output to one input of a 7400 NAND gate. The other input is connected to a normally off clock pulser,

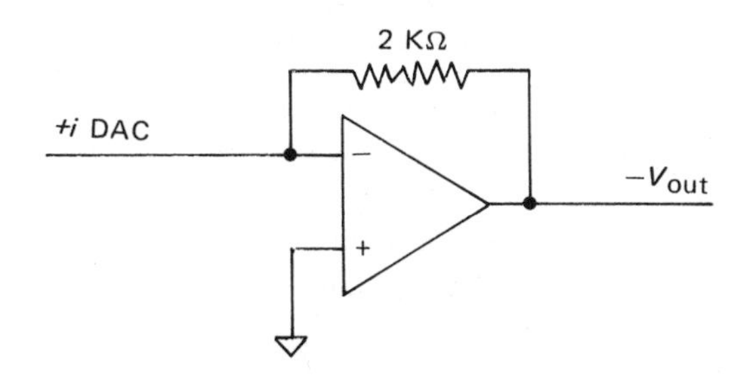

Fig. 5-10
Current-to-voltage converter.

which had previously been connected directly to the 7493 counter. The gate output goes to the input of the 7493 counter. At this point, the ADC is completed. To check its operation, reset the counter, and set -3 V on the voltage reference source (VRS). Make sure the Q output of the flip-flop is at logical 1. Now pulse the counter. It should count up only to that value corresponding to -3 V and stop. Continued pulsing is prevented from reaching the counter through the NAND gate as long as the flip-flop remains at logical 0. During this time, the contents of the counter can be strobed into a data latch which will store the data even after the counter is reset. Clearing the counter, then setting the flip-flop, will allow repetitive runs.

6. Connect power and common to the 7475 data latch according to the data sheet. Use decoupling capacitors. Wire the A, B, C, and D outputs of the counter to the four latch input pins. Connect the Q outputs of the latch to indicator lights. Make sure that you keep the A, B, C, and D sequence. Take the signal coming from the RS flip-flop, invert it using a 7404 gate, and connect it to the two clock inputs of the latch. Only when the flip-flop's Q output changes to 0 will data be strobed into the latch. If the set input of the flip-flop is held at ground, the counting can be resumed without the contents of the latch changing from that when the counting was first inhibited. Plot the latch contents as a function of different VRS voltages from 0 to -5 V to examine how well your analog-to-digital converter will perform. How good is its resolution? If you modify the circuit to provide automatic repetitive conversions, be sure to check your proposed circuit with the laboratory instructor before proceeding to wire it. What is the conversion time of the ADC for the maximum voltage? Minimum voltage?

C. Analog-to-Digital Converter (ADC): Continuous-Ramp Voltage-Comparison Converter (Optional)

A continuous-ramp voltage-comparison converter can be constructed by replacing the counter-DAC portion with an operational amplifier in the ramp configuration.

1. Wire the circuit shown in Fig. 5-11. The ramp voltage at the output of the 72741 integrating amplifier will vary from 0 to -5 V over a period of about 10 sec since

$$V_0 = -1/RC\int_0^t V_{\text{in}}\,dt \qquad (5\text{-}4)$$

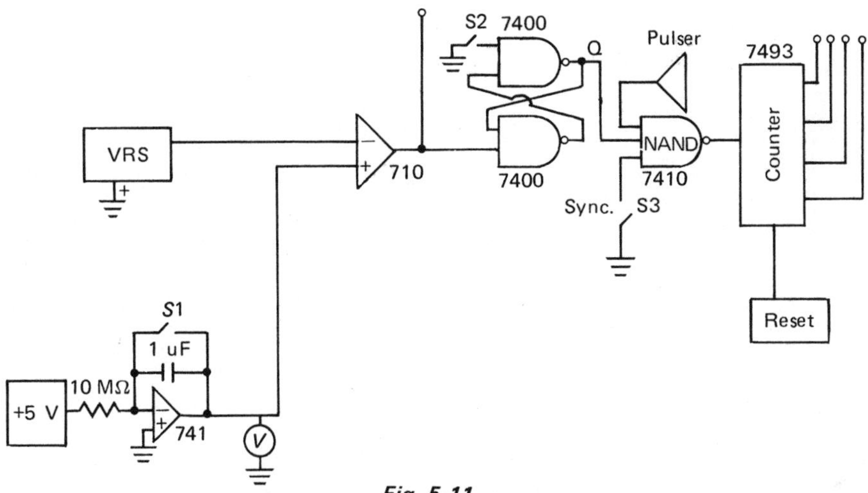

Fig. 5-11

Continuous-ramp voltage-comparison converter.

Use the +5 V on the EL Digi-designer or a +5-V power supply for V_{in} to the integrator. Monitor the ramp output with the VOM. Do not let the ramp voltage go more negative than −7 V or the comparator will be damaged. The ramp comparator can be conveniently shorted, using a toggle switch connected across the capacitor. Familiarize yourself with this component before attaching it to the comparator. Use VRS values between 0 and −5 V.

2. To initiate a run, clear the counter, set the VRS (−) voltage, set the RS flip-flop by momentarily closing *S*2, and then open the capacitor switch (*S*1) and the sync. switch (*S*3) simultaneously. At the end of a run, close the sync. switch and capacitor switch. The sync. switch, while closed, will keep clock pulses from reaching the counter. Once open, only the control flip-flop can disable the counter. The actual readings you obtain for a given VRS signal will depend on the frequency of the pulse generator, which must be kept constant for a given series of runs. For example, setting the pulse frequency control at about 1 Hz, a VRS signal of −4 V results in a count of about 7. Plot counter output vs VRS voltage at two or more clock frequencies. Note the long conversion time.

D. Continuous-Balance Voltage-Comparison ADC

Figure 5-12 shows the completed circuit you will wire.

1. Wire the 74193 up/down counter. Pins 7, 6, 2, and 3 are the counter outputs. Monitor them with lights and connect them to the four

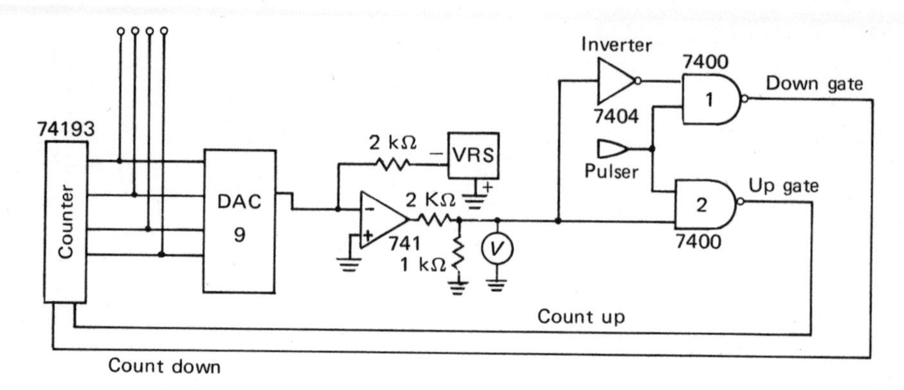

Fig. 5-12
Continuous-balance voltage-comparison converter.

most significant DAC bits. Check the operation of the counter by connecting pins 4 and 5 to separate, normally on pulsers. Pulsing pin 4 causes the chip to count down, while pulsing pin 5 causes the chip to count up. Attach a normally off pulser to pin 14. This may be used to clear the counter. Refer to the data sheet.

2. Wire an operational amplifier in the open-loop configuration (no feedback resistor), as shown in Fig. 5-12. In this configuration, the amplifier functions as a comparator. Again use the ± 15 V power supply for the DAC and operational amplifier, using decoupling capacitors as needed across ± 15 V and ground. The VRS negative lead is connected to one 2-kΩ resistor, and this component into the (−) input of the operational amplifier. The positive lead is connected to ground. This unit will then generate the required negative voltages between 0 and -5 V. When the VRS is set to -5 V, it will generate -2.5 mA, which should match the maximum output current of the DAC. In practice, it may be necessary to set the VRS as high as -7 V to balance the maximum DAC output. The DAC output is also connected to the (−) input of the 741. Check the comparator's operation. Set -4 V on the VRS and reset the counter. With a normally on pulser connected to pin 5 of the counter and pin 4 left open, the counter will increase as one pulses it. Initially, the comparator should be logical 1 ($+15$ V). (The voltage divider network to the right of the amplifier acts to reduce the output to a TTL comparable ± 5 V.) At a count of about 8 or 9, the comparator output should change to 0 (-15 V). Plot the count at which the comparator changes state for initial VRS settings of -1, -2, -3, -4, and -5 V.

3. Now connect the comparator as shown in Fig. 5-12. Use a normally off pulser connected to gates 1 and 2. Connect the down gate output to pin 4 of the counter and the up gate output to pin 5. While continuously pulsing the counter, note how the VRS signal is followed by the digital output as it is varied between 0 and −5 V.
4. Substitute a pulse generator set at about 1 Hz for the hand pulser. Include a data latch to store the counter value at regular intervals, and wire as in Fig. 5-9 and Sec. A. This circuit corresponds closely to a commercial ADC. Design the modified circuit and check with the lab instructor before implementation. Set various values on the VRS, and monitor the data latch after the counter output has been strobed in.

Section **II**

Experiments 6–17: Principles of Interfacing

Introduction to Experiments 6 – 17: Principles of Interfacing

The experiments in this section are designed to introduce the applications of digital computers in the laboratory. The first experiments serve as a bridge between the digital logic experiments of the preceding section and the more advanced data-acquisition and control experiments which follow them. The later experiments are built upon a foundation of digital logic and computer-interfacing principles and provide demonstrations of actual laboratory application of such techniques. The student is required to understand and execute chemistry laboratory experiments as usual—but, in addition, he is required to design the appropriate programming and communication elements for optimum interaction between the computer, experiment, and experimentalist. Thus, the student's achievement is one of *experimental design* and the *proper utilization* of computing equipment in the solution of measurement problems in chemistry.

The important prerequisites of the laboratory experiments are: (1) a modified high-level programming language, which includes versatile data-acquisition and experimental-control subroutines with conversational mode calling sequences; and (2) a general-purpose hardware package for interface design. This hardware should provide for patch-board incorporation of those digital and analog modules required to complete an interface between computer and experiment. Such a general-purpose interface package requires only that the student be able to lay out the basic logic design, timing sequence, and the analog amplification or attenuation required. He can then select appropriate elements to implement his design.

REAL-TIME BASIC SYSTEM

The high-level programming language chosen for use is the BASIC language. There are several reasons for selecting this language: (1) It is *easy to learn*, generally requiring about half a day's exposure to develop a working knowledge. (2) It is an algebraically oriented *conversational* language. (3) It is *interactive.* That is, BASIC compilers are usually interpretive and therefore compile and execute programs line-by-line. This allows programs to be entered, executed, and edited on-line through a teletype terminal. It also provides for immediate turn-around and rapid error diagnostics. (4) BASIC is rapidly becoming a *universally acceptable* language. (5) BASIC is currently readily *available* at commercial time-share terminals and on most laboratory computers.

The software system will be referred to as "REAL-TIME BASIC" (RTB). The software includes, fundamentally, a BASIC interpreter with machine-language subroutines which are directly callable from the BASIC software and which are designed to communicate in a variety of ways with experimental systems. A typical RTB data-acquisition and control software package is described below.

GENERAL DESCRIPTION OF RTB SYSTEM

Figure II-1 illustrates the general nature of an on-line computer system. On the one hand is located the laboratory experiment, the source of data which it is desired to present to the computer for data processing. On the other hand is the digital computer itself, which is capable of high-speed

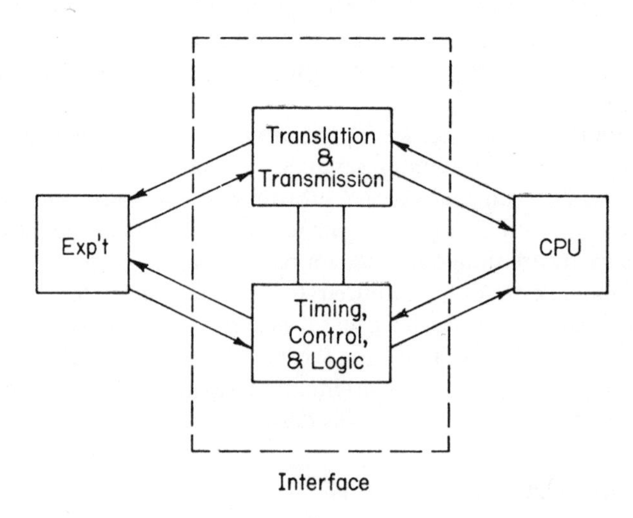

Fig. II-1

Schematic diagram of on-line computer configuration.
[From *J. Chromat. Sci*, 7, 714 (1969).]

programmed computational and control functions. In order to realize a communication link between the computer and the experimental system, an electronic interface must be established. This interface accomplishes the functions of logical control, synchronization, timing, and translation (analog-to-digital, digital-to-analog, decoding, logic conversions, etc.). The programming language must take into account the nature of the electronic interfacing, and the interfacing must be designed with the fundamental characteristics and capabilities of the programming language in mind. Similarly, the subroutine package must take into account general features of the hardware interface module. Therefore, before describing the general-purpose subroutine package, it will be necessary to give a general description of the interface.

GENERAL PURPOSE INTERFACE HARDWARE

Specific design details for one possible interface module are given in Appendix F. However, it is expected that whatever interface is used will have the following features:

1. A series of control (sometimes called ENCODE) flip-flops which the central processing unit (CPU) can use to alter the logical status of external devices. These *one-bit logic* modules should be capable of delivering the current necessary to drive relays, etc., and should be resettable by the CPU.
2. A complementary series of sense line (FLAG) flip-flops through which the CPU can ascertain the logical state of an attached line, and which can also be reset by the CPU.
3. At least one register, containing as many bits as there are in one computer word. This register should be designed so that it may be set and read either by the CPU, or by an external device. In addition to this, it will be assumed that either the console itself or the interface contains a series of switches which can be used to directly enter data into the computer during program execution.
4. A programmable clock, capable of running at several rates, which can be set, read, started, and stopped.
5. An analog-to-digital converter (ADC) with at least a two-channel multiplexer, if possible. At lowest gain, the ADC should have an input range of about ± 10 V.
6. At least one digital-to-analog converter (DAC) capable of converting 10 bits or more and with a nominal output range of ± 10 V. Two such converters are highly desirable.

7. The devices enumerated in 1–6 above should, ideally, have inputs and outputs readily accessible *via* an external patchcord panel. (Although this is not strictly necessary, it greatly facilitates wiring of experiments to the general-purpose interface.)

The ENCODE and FLAG flip-flops themselves are commonly just individual bits of two special registers. Therefore, for the purposes of this discussion, we shall assume that there are as many control and sense bits each as there are bits in the computer word. In addition, device types 3–6 will, in general, require ENCODE control pulses to start them and FLAG sense lines to detect when they are ready. Therefore, one ENCODE–FLAG bit pair should be allowed for each such device; it is still possible, however, to use the ENCODE signal to control the simultaneous start of one of these devices and one or more external devices. Each device, and its associated ENCODE–FLAG bits, will be identified by a unique device number which is computer selectable.

REAL-TIME BASIC SUBROUTINES

We will now consider the types of assembly-language subroutines we have added to BASIC to augment it, and yield that version we call REAL-TIME BASIC. It is well to remember that there are certain restrictions implicit in the use of an interpretive compiler, such as this, in laboratory computing. The amount of memory available will form one restriction. In addition, there are certain time constraints. Since the edited source code must, for most laboratory BASIC compilers, be reinterpreted each time it is executed, the speed with which programs may be run will be limited by this fact. In practical terms, if data are being gathered by repeated calls, to ADC (see below), then a maximum collection rate of perhaps 30 to 100 points per second can be realized. The exact rate will depend, of course, on the number of statements contained in the data-collection loop. High rates may be achieved by collecting all the data in one block within an assembly-language subroutine, without returning to BASIC until data gathering is complete. These are limitations which, for the most part, are minor and which can be overcome in the ways we will suggest.

As was mentioned earlier, laboratory computing requires an easily definable set of subroutines which are needed to exercise the special data acquisition and control peripherals which are used. Since the specific details of these devices vary from laboratory to laboratory, it is probably not useful to give such details in this section. (However, details for implementing RTB on two specific computers are given in Appendices A and B.) Instead, we shall attempt to define the *functions* which are required in the laboratory and a methodology. One convenient categorization of required functions is as follows:

Control pulses, relay closures, and senses
Digital input and output
Timing and synchronization
Analog input
Analog output

Various versions of RTB, depending on the experiment requirements and the array of peripheral devices, are therefore possible. A summary of one such set of subroutines and their functional characteristics is given in Table II-1.

Table II-1

A typical REAL-TIME BASIC Subroutine Package

Subroutine	Function
CNTRL(*C*)	Sets the ENCODE bit associated with device "*C*"* and waits for the FLAG associated with device "*C*" before continuing. FLAG and ENCODE bits "*C*" are cleared (reset) before exit.
STATUS(*C*)	Waits for the FLAG bit "*C*" before continuing. FLAG bit "*C*" is reset before exit.
WAIT(*S*)	If the most significant bit of the console (interface) switches is zero, the binary number entered on these switches is read, converted to decimal, and stored in variable "*S*." If the most significant bit is set, the computer waits until it is turned off, then returns. "*S*" must be specified in the CALL statement in *either* case.
INBIT(*Z*, *S*, *C*)	Sets ENCODE bit "*C*" and waits for FLAG "*C*" before inputting status of bit "*Z*" from device "*C*." Value of "*S*" is made "1" or "0" accordingly. FLAG and ENCODE bits are reset before exit.
OBIT(*Z*, *C*)	Causes the CPU to output a "1" to bit "*Z*" on logical device "*C*" (normally a register). ENCODE and FLAG bits are unaltered.
DIGIN(*Z*, *C*)	Sets ENCODE and waits for FLAG before inputting the contents of register "*C*." The floating-point equivalent of the register contents is saved in variable "*Z*," and ENCODE and FLAG are reset before exit.
DIGOUT(*Z*, *C*)	Sets ENCODE and causes the binary equivalent of decimal number "*Z*" to be output to device "*C*," after which it waits for FLAG. FLAG and ENCODE are reset before exit.
INIT(*F*)	Selects the clock frequency. "*F*" is the desired frequency for the clock. ENCODE and FLAG bits on all channels are unaltered. *This statement must precede any call to* START.
START	Used to start the clock. The ENCODE bit associated with the clock must be connected to the clock start line. Alternately, the encode bit may be connected to some external logic circuit which can be used to synchronize the clock start with an external event such as the start of the experiment (see the discussion). This command also clears the ADC FLAG, and starts the ADC by setting its ENCODE bit.

*Input/output channels and device codes are usually designated as octal numbers for laboratory computers. Since BASIC compilers normally accept only decimal numbers as input, the codes and channels here are specified in decimal.

Table II-1 (continued)

Subroutine	Function
STOP	Turns off the clock by clearing the ENCODE and FLAG bits. This can be used to stop experiments, *etc.*, which are connected to the clock ENCODE bit. The ADC ENCODE bit is also cleared to reset the ADC.
ADC(X)	Waits for ADC FLAG, indicating that conversion is complete and that the preselected time interval is up. Then, one datum is taken from the ADC, converted to millivolts in floating point, and saved in variable "X." The clock and ADC are then cleared and restarted. FLAG is cleared; ENCODE is unchanged.
BLOCK[X(1), F, T]	Takes in a complete block of "T" points, which are stored sequentially in array "X." "F" determines the clock frequency (as in INIT). Data are saved as millivolt values. The data-acquisition time base can be synchronized with the start of the experiment through the external clock control line. FLAG and ENCODE bits are cleared before exit.
DAC(D)	Causes the output of "D" millivolts from the DAC. If more than one DAC is available, an additional argument specifying which one the routine refers to must be added. Timing for repetitive outputs may be achieved by using START and either CNTRL or STATUS, where the output of the clock is wired to the appropriate FLAG.

Using these routines, all the experiments in this part of the manual may be performed. This particular set is a slight modification of packages used with the Hewlett-Packard 2100 series computers at Purdue University and Varian 620 series computers used at the Universities of Nebraska and Oregon. Since the detailed nature of any such package is dictated, to a certain degree, by the specific details of the computer and interface hardware used, *minor differences* in the number and kind of subroutines required to perform *functionally identical operations* with various different laboratory computer systems may occur. Accordingly, it is likely that the RTB subroutines the reader will face will be yet another version of REAL-TIME BASIC. For instance, in some cases, subsets of the routines must be used due to memory limitations. This is, of course, a trivial problem since once a package of assembly language routines of this sort is operational, reassembly to fit changing requirements is simple.

Nevertheless, the subroutines given here will illustrate the general requirements of RTB. The reader should translate the examples given in the following pages into the specific dialect of RTB available for his computer system. Thereafter, he should experience little difficulty writing programs for the subsequent experiments in this section.

Since we do not intend to provide an exposition of BASIC programming in this text, it is assumed that the student has mastered the fundamentals of BASIC before attempting the experiments outlined in this section. (The Bibliography lists several excellent texts which teach fundamentals of BASIC.) The use of subroutines such as those described here is a relatively simple matter requiring only that the student have a good understanding of the functions of each subroutine and that he be able to insert the appropriate CALL statements at logical places in his programs.

CONTROL PULSES, SENSES, AND DIGITAL I/O

One of the principal requirements for a laboratory computer is that it be able to control external devices and to detect logical changes in these external devices. One subroutine available for this kind of operation is CNTRL. CNTRL causes the output of an ENCODE bit on a selected I/O channel to change state (go to the "TRUE" level) when CNTRL is called. The computer then waits within CNTRL for an external event to occur which will cause a FLAG bit on the same I/O channel to change to a TRUE state. The subroutine detects this event and then allows the next sequential program statement to be executed. This next statement might conceivably be a call to a data-acquisition subroutine. STATUS does not cause a change in the ENCODE bit, but otherwise operates the same way. Thus, if control and detection are necessary, CNTRL should be used; but if only detection is desired, STATUS is sufficient.

A simpler, but less precise, means of communication with the computer is by WAIT. When this subroutine is entered, the computer simply waits until the operator flips a toggle switch corresponding to the most significant bit on the computer console switch register. The computer then exits WAIT and executes the next program statement—which might be to call START to begin data acquisition. WAIT can also be used to enter binary data directly into the computer.

Subroutines OBIT and DIGOUT allow the programmer to utilize specific *output bits* on a selected I/O channel to control external devices. There are as many output bits available for this function as there are bits in the laboratory computer word. With OBIT, the user can select which bit he wants to set by specification of the bit number. The setting of one of these output bits causes a corresponding binary voltage level change at the specified output terminal, and this can be used to close or open switches, start or stop experimental events, light indicator lamps, *etc.* DIGOUT allows the programmer to output a one-computer-word binary voltage *pattern* with any simultaneous combination of logical 1's and 0's he chooses. This binary pattern is selected by including in the subroutine call the decimal equivalent of the binary number to be generated. Thus, more than one event can be controlled simultaneously.

INBIT and DIGIN provide digital *input* information for program "sensing" of external situations. Thus, many binary voltage input terminals are available to the user, the status of which are acquired by either subroutine.

INBIT transmits to the BASIC program the status of a single specified input bit. This subroutine is useful to the student because he need only specify a bit number in the call statement to check the status of a bit. DIGIN is similar to INBIT, except that it transmits to the BASIC program the numerical equivalent of the input, a computer-word-long binary voltage pattern. Because the status of these bits can be controlled by external events, the computer could use this information to make appropriate changes in the data processing or control programming. Of course, WAIT can be used if the programmer desires to control the program changes personally.

TIMING AND SYNCHRONIZATION

The most fundamental operation which must be accomplished by the RTB system is the generation of a *time base* for all experimental functions. This is accomplished by the incorporation of a fixed crystal clock into the interface hardware. It is also necessary to include electronic count-down logic to scale the output clock pulses down to a usable frequency range for chemical experimentation (100 kHz to 0.01 Hz). The countdown logic should be designed in such a way that various subfrequencies may be stipulated by the program. The programmed clock output pulse train is then available to control the timing on the ADC, DAC, or any other external hardware. Also available simultaneously may be synchronous clock pulses representing frequencies at the various stages of countdown.

The clock is controlled by the RTB software through subroutines INIT, START, and STOP. INIT is the *initialization* subroutine. Through it the programmer specifies the clock frequency, within the limits outlined above.

START is called when the computer program decides that it is time to *enable* the clock. That is, an ENCODE bit is set to a "1" state on the data-acquisition channel. This bit can be connected externally with a patchcord to the ENABLE terminal on the clock module (see Fig. II-2a). Alternatively, the ENCODE bit may be brought to some external logic circuit, the output of which will set the ENABLE clock input when other external events have occurred, such as the start of an experiment. One possible external logic configuration is shown in Fig. II-2b, where the ENCODE bit conditions one input of an AND gate. The other input is conditioned "TRUE" when the experiment has been initiated. The AND gate output will then enable the clock. At that point the countdown logic will be enabled and programmed clock pulses will begin to appear at the available outputs.

STOP is called to disable the clock when the time base is no longer required. It simply "clears" the ENCODE bit on the data-acquisition channel which has been used externally to enable the clock.

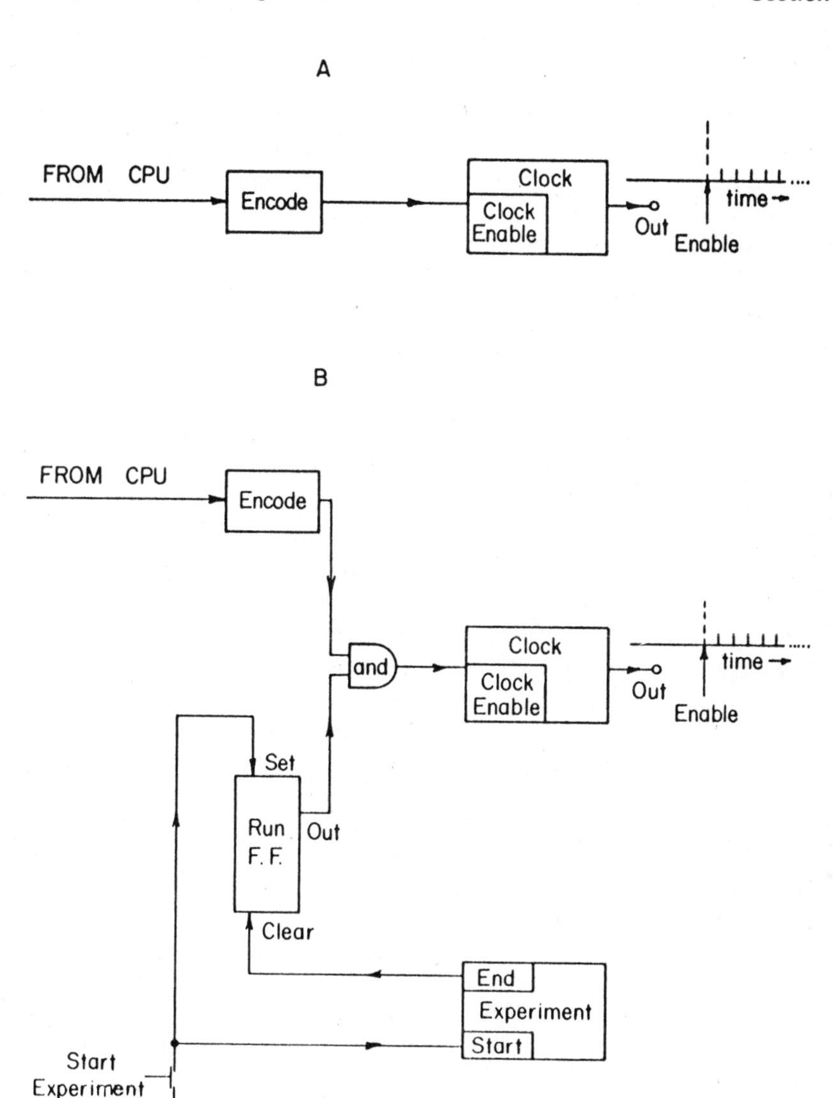

Fig. II-2

Synchronization of clock output with computer and experimental events. (a) Direct computer control of clock enable input. (b) Clock enable controlled by AND gate output. Clock enabled when both computer command and experimental start are seen at AND gate inputs. The "RUN" flip-flop output follows the status of the experiment; it is set "true" when the experiment starts and "false" when the experiment ends. Thus, the clock is disabled when the experiment ends. [From *J. Chem. Educ.*, **48**, 317 (1971).]

Another type of communication required between the experiment and computer is associated with *synchronization* between computer operations and external events. For example, if the computer has completed all preliminary program execution required before being able to accept experimental data, the user may choose to receive an output from the computer which may not only tell the experiment to start, but may also initiate external events associated with the experimental system. Clearly, several of the subroutines already discussed have this capability.

CNTRL, STATUS, INBIT, and DIGIN can all be used to sense the conditions associated with an external environment and can therefore be used for synchronization. CNTRL and STATUS use the FLAG status bits directly, waiting until they are true, then exiting to execute the next BASIC statement. INBIT and DIGIN return directly to BASIC and require additional program code to provide *software* synchronization, but are slightly more versatile because the programmer may choose to synchronize on either true or false states, and because DIGIN can synchronize with a rather large combination of states (which would have to be hardware AND'ed to use CNTRL or STATUS). On the other hand, the BASIC code used to implement INBIT and DIGIN will be somewhat slower than that used by STATUS, and the synchronization will be somewhat less precise. Similarly, WAIT can be used if synchronization is to be under human control. We have previously shown how the clock can be wired to synchronize with the start of an experiment. Finally, the data-acquisition routines discussed in the next section provide several means of synchronization.

DATA ACQUISITION

Two different types of data-acquisition subroutines are included in the package summarized in Table II-1. They both work in conjunction with an analog-to-digital converter (ADC) that is controlled by the external clock. The first mode of acquisition allows one data point at a time to be digitized and stored, before returning to the BASIC program from the subroutine. The second mode provides a means of acquiring an entire block of data (a series of data points which are digitized and stored) before returning to the main program.

To collect one data point at a time, the user simply inserts the statement CALL INIT, followed by the CALL START and CALL ADC statements at the appropriate point in his program. This causes the computer to monitor the analog-to-digital converter's "conversion complete" FLAG, indicating the current datum has been digitized. When this happens, ADC takes the data point from the converter, converts the datum to floating-point format, and stores it in the appropriate memory location for subsequent reference by the BASIC program. When this action is complete, the subroutine returns control to the next statement in the BASIC program. The newly acquired datum can then be operated upon. In order to acquire another data point, the sequence must be repeated. A series of points may be accumulated in this way by including the CALL ADC statement in a loop. (INIT need be called only once, and ADC restarts the ADC, so that repeated calls to START are not necessary.)

A second type of data-acquisition subroutine (BLOCK) is one which allows the acquisition of a complete block of data before exit. The external clock is started and synchronized with data acquisition within the subroutine. The basic difference between BLOCK and ADC is that ADC is called to acquire one data point at a time and therefore allows for BASIC program statements to be executed during the time between acquisition of data points. Thus, an experimenter could devise a program which processes experimental data while the experiment is in progress and data are being acquired. This is referred to as "Real-Time" data processing. However, because the computer program must involve the relatively inefficient execution of BASIC statements between data points, there is a more severe limit on the speed with which data can be acquired without the computer getting out of synchronization with the experimental timing (i.e., without more than the preselected clock time passing between data-point acquisitions). If this happens, an error message is printed, and the computer halts. In this case, the user must revise his program to require less real-time processing. With BLOCK, on the other hand, a complete block of data points is acquired before the subroutine is exited and control returned to BASIC. Therefore, the timing is limited by the efficiency of the machine-language programming comprising BLOCK and by the ADC conversion time. Typically, using the Purdue hardware (Appendix F), it is possible to acquire data at rates as great as 20 kHz with BLOCK, but no real-time data processing is possible.

The limiting data rate when using ADC is about 30 to 100 Hz, with a minimum of real-time data handling. If several BASIC computations are to be executed between data points when using ADC, the limiting data-acquisition rate may be on the order of 1 to 10 Hz. However, this is generally more than adequate for most experiments in the chemistry laboratory.

ANALOG OUTPUT

In addition to performing data acquisition, there are several other ways in which the computer can communicate with an experiment. One of these is through a digital-to-analog converter (DAC), where the digital output of the computer is converted by the DAC to an analog voltage level. One of the

RTB subroutines, DAC, provides the capability of driving the DAC, the output of which can be connected to external experiments. It is possible to generate a virtually continuous voltage waveform output from the DAC by mathematical generation within BASIC of the discrete points making up the waveform and transmitting these through subroutine DAC to the digital-to-analog converter in a repetitive fashion synchronized with an external clock.

DATA ACQUISITION AND GENERAL-PURPOSE INTERFACE HARDWARE

A schematic diagram of a general data-acquisition system operated in conjunction with RTB is shown in Fig. II-3. The system shown here is designed around a computer with a word length of 16 bits. The I/O bus is a

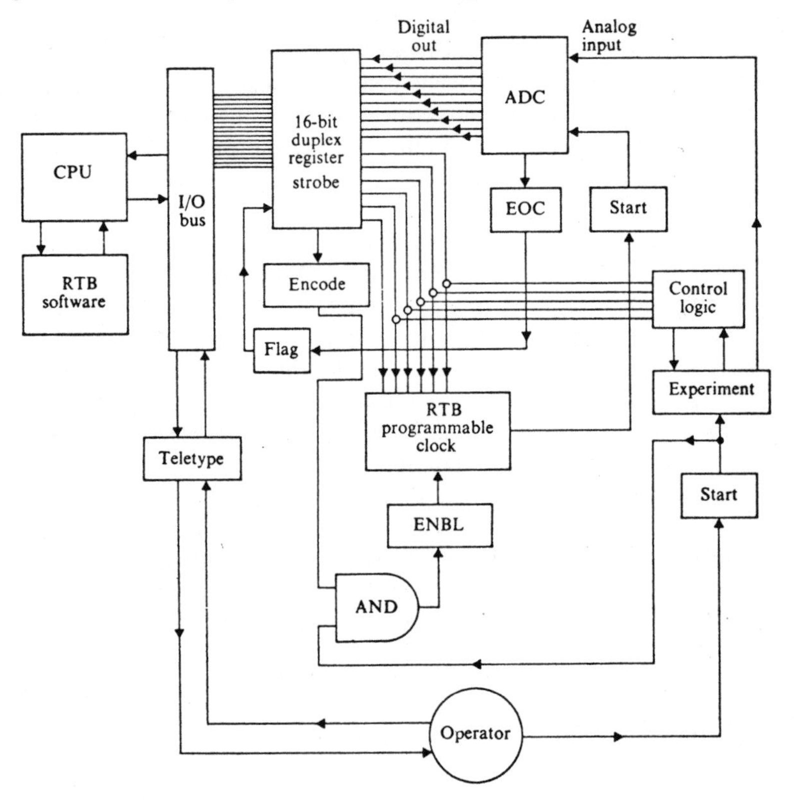

Fig. II-3

Typical REAL-TIME BASIC experimental data-acquisition setup.
[From *J. Chem. Educ.*, **48**, 317 (1971).]

set of wires carrying all the computer input and output signals. This has been jumpered to the interface. The programmable clock has been wired so that it can be started synchronously with an experiment. (See Appendix F for details.)

EXAMPLES OF USE OF RTB SYSTEM

Example 1. Take in 100 digitized data points, at a rate of 1 per second. Sum those which are greater than +10 mV. Synchronize the start of data acquisition with setting of switch 15 on the console.

	BASIC Program	Comments
1Ø	LET X = Ø	Define data variable
2Ø	LET F = 1	Set frequency = 1 Hz
3Ø	LET S = N = Ø	Initialize sum and counter
5Ø	CALL INIT(F)	Initialize CLOCK
6Ø	CALL WAIT(Z)	Wait for switch closure. Sw 15 must be set before RUN
7Ø	CALL START	Start CLOCK
8Ø	CALL ADC(X)	Wait for and take ADC datum
9Ø	IF X ≤ 1Ø THEN 11Ø	Check new *X*
1ØØ	LET S = S + X	Add value > 10 mV to sum
11Ø	LET N = N + 1	Increment counter
12Ø	If N < 1ØØ THEN 8Ø	
13Ø	PRINT S	
14Ø	END	

Several aspects of the program above should be discussed. First, the execution of START (statement 70) takes place within a short, but finite, time after the switch 15 closure is recognized by WAIT (statement 60). This time delay is imposed because the next BASIC statement must be accessed before moving on. Perhaps 1 msec transpires before the CLOCK is started by START. Thus, the synchronization of the CLOCK with the start of the experiment may be in error by this amount. For data-acquisition rates up to about 50 or 100 Hz, this error is insignificant, but it becomes serious for higher data rates. Thus, for higher data-acquisition rates, an alternative synchronization approach, such as that shown in Fig. II-2b, should be used.

A second aspect of Example 1 is that in ADC (statement 80) the computer waits for the right time to take the next data point. Thus, it is called immediately after starting the CLOCK, and the computer waits for about 1 sec for the first data point. After the first data point is acquired, it can be processed, as *X*, until the next data point becomes available. At this time, the computer must have completed all processing and be ready for the next data point.

In other words, program control should have been returned to statement 80. Thus, for this example, execution of statements 90, 100, 110, and 120 must be complete within 1 sec. This is a reasonable expectation since each BASIC statement is accessed and executed in a time on the order of between 0.5 and 10 msec. (Longer times may be required for complex arithmetic statements.) If, however, the data acquisition rate is set higher, or more processing is attempted between data points, program control may not return to statement 80 before the next data point is available. If this should occur, an error routine will cause an error statement to be printed, and the computer will halt.

It should also be noted that the start of data acquisition could have been synchronized with an external electronic event by using START or CNTRL instead of WAIT (statement 60).

Example 2. Acquire a block of 200 digitized data points from an experiment at a rate of 1 kHz. Process the data as in Example 1 after acquisition is complete.

```
 2Ø    DIM X(2ØØ)
 3Ø    LET S = Ø
 4Ø    LET F = 1ØØØ
 5Ø    LET T = 2ØØ
 7Ø    CALL BLOCK (X(1), F, T)
 8Ø    LET I = 1
 9Ø    IF X(I) ≤ 1Ø THEN 11Ø
1ØØ    LET S = S + X(I)
11Ø    LET I = I + 1
12Ø    IF I ≤ T THEN 9Ø
13Ø    PRINT S
14Ø    END
```

It should be noted that in Example 2 BLOCK was called without calling the initialization routines INIT and START. This is possible because BLOCK incorporates the functions of both INIT and START; that is, BLOCK initializes the data array, interprets the frequency request, starts the clock, takes in the data, and converts it to millivolts (floating point) before returning control to to the next BASIC statement.

Example 3. Take 100 data points at variable rates: first 10 points at 10 Hz, next 90 points at 1 Hz. Synchronize the start of data acquisition with an external event.

```
2Ø    DIM D(1ØØ)
3Ø    LET X = Ø
4Ø    LET I = 1
5Ø    LET F = 1Ø
```

```
 6Ø     CALL INIT(F)
 7Ø     CALL CNTRL(1)
 8Ø     CALL START
 9Ø     CALL ADC(X)
1ØØ     LET D(I) = X
11Ø     LET I = I + 1
12Ø     IF I#1Ø THEN 15Ø
13Ø     LET F = 1
14Ø     CALL INIT(F)
15Ø     IF I ≤ 1ØØ THEN 9Ø
16Ø     (PROCESS  DATA)
        .
        .
        .
5ØØ     END
```

Note that the clock frequency does not change, in the above example, until *after* the tenth datum has been acquired, even though INIT was called between the ninth and tenth data samplings. This is an important feature of the RTB clock. The frequency cannot be changed during a clock period.

Alternatively, the program for Example 2 could be written using BLOCK as long as no data processing is required between data points:

```
 2Ø     DIM X(1ØØ)
 3Ø     LET T = 1Ø
 4Ø     LET F = 1Ø
 6Ø     CALL BLOCK(X(1), F, T)
 9Ø     CALL BLOCK(X(11), 1, 9Ø)
11Ø     (PROCESS  DATA)
        .
        .
        .
5ØØ     END
```

Note that statement 90 refers to the 11th location in the data block assigned to the parameter X. This is because the first 10 locations have previously been filled with the values of the first 10 data points.

These two programs would be useful for taking data from an experiment where the data were changing rapidly at first, but where the rate of change slowed as the initial stages of the experiment passed. Neither program would be appropriate for very high data rates because of the timing errors introduced when changing clock frequencies. When using ADC, however, no timing error is introduced when changing frequencies in real time *unless* the time required to change the frequency exceeds one time period of the prior clock frequency. If the time required *is* excessive, ADC will print an error statement, and the program will halt.

Example 4. Use the digital-to-analog converter to output a ramp voltage varying from 0 to +5.00 V at a rate of 100 mV/sec in 10-mV steps.

```
 2Ø    LET D = Ø
 3Ø    LET F = 1Ø
 4Ø    CALL INIT(F)
 5Ø    CALL START
 6Ø    CALL STATUS(7)
 7Ø    CALL DAC(D)
 9Ø    LET D = D + 1Ø
1ØØ    IF D ≤ 5ØØØ THEN 6Ø
15Ø    END
```

Here, the CLOCK FLAG (time-up) has been assigned device address 7. How would you write the above program for 1-mV steps?

DIGOUT could be used to control an external multiplexer that runs on a simple binary counter. To sample four channels, for example, one would program:

```
  1    LET C = 1
  5    LET X = 1
 1Ø    LET F = 1Ø
 2Ø    DIM D(4)
 3Ø    CALL INIT(F)
 4Ø    CALL START
 5Ø    FOR I = 1 TO 4
 6Ø    CALL DIGOUT(I, C)
 7Ø    CALL ADC(X)
 75    LET D(I) = X
 8Ø    NEXT I
 9Ø    CALL STOP
        .
        .
        .
2ØØ    END
```

DIGIN could be used to bring in an external counter value and would be programmed as follows:

```
 5Ø    LET C = 1
 6Ø    CALL DIGIN(Z, C)
 7Ø    PRINT Z
        .
        .
        .
```

INBIT could be used in program branching as follows:

```
 1Ø    LET Z = 1
 2Ø    LET C = 8
 3Ø    CALL INBIT(Z, S, C)
 4Ø    IF S#1 THEN 3Ø
 5Ø     .
 6Ø     .
        .
        .
```

CONCLUSION

To summarize, the REAL-TIME BASIC approach to computer interfacing is simple in concept. A general-purpose hardware interface for common data acquisition and control functions is used with a standard BASIC compiler, to which have been added the requisite assembly-language subroutines for control and monitoring of the general-purpose interface. The RTB software package thus facilitates the programming aspects of the interface problem. Since system dependence has purposely been minimized in the discussion above, the hardware/software package described does not correspond in exact detail to any in existence. Appendices A and B contain explicit descriptions of the actual RTB systems used in the development and testing of the experiments in this section. In any case, a system possessing the *functional capabilities* described above will be completely adequate for performance of the experiments which follow.

Experiment 6

Computer-Controlled Logic Chip Test

In experiment 1 the truth tables of several logic chips were determined. Although the procedure used to test these chips is relatively simple, it would not be feasible to use it if a large number of chips were to be tested; the process of hand-testing each chip would rapidly become too tedious for even the most patient worker! Because computers never become bored, small computers are frequently used to perform repetitious tasks of this nature; thus the computer-controlled test of logic gates provides an ideal example for the first interface experiment. Other common applications which will be included in later experiments include high-speed data acquisition, manipulation, and presentation; signal recovery by averaging methods; and the simultaneous control or coordination of several different experiments.

The purpose of this experiment is to acquaint the student with the interface capabilities of a laboratory computer and to demonstrate how such a computer can detect and control external phenomena (in this case the logic levels of a chip). It is assumed that the laboratory computer available has been equipped with:

1. A general-purpose register which can be gated to either transmit its contents to the external environment or to receive information from an external device.
2. Several 1-bit registers (sense lines) which can detect the logical states of attached external lines. Moreover, for the purpose of this discussion, the general-purpose register is assumed to use *negative* logic; i.e., logic 1 is 0 V, while logic 0 is +5 V.

Before designing the interface, it is necessary to devote some thought to the programming problems involved. One problem is that of varying the chip inputs so that all combinations are tested, and of doing so in a systematic manner to assure that none are missed. The different logic conventions used in the register and chips pose a second problem which must be considered.

Table 6-1

Input Combinations for a 3-Input Gate

Decimal value	Input 1	Input 2	Input 3
0	0	0	0
1	0	0	1
2	0	1	0
3	0	1	1
4	1	0	0
5	1	0	1
6	1	1	0
7	1	1	1

The output of the tested gate must be detected, and a method found to determine whether the gate is functioning properly. Finally, if more than one type of gate is tested, a method must be found to automatically determine the type of gate.

A moment's reflection shows that one way to vary the gate inputs to generate a truth table is to use the binary-number system. Thus, for a three-input gate, all possible input combinations are represented by counting in binary from 0 to 7, as shown in Table 6-1.

The second problem is most easily solved by inverting the output of the register. Inverting the output has the effect of converting the negative logic to positive logic. It is also possible to solve this problem by using negative logic in the program; i.e., treat logical true as a zero, and logical false as a one. This latter approach is left as an exercise for the reader.

The sense line provides a means of detecting the logic gate output. Detecting inoperative gates and determining the gate type are actually different sides of the same problem, and the solution depends on realizing that each type of gate generates a different output pattern if its two inputs are varied in the same systematic way (Table 6-2). Thus, when the input conditions are varied by counting from 0 to 3 in binary, the output pattern is 0001 for an AND gate, 0111 for an OR gate, 1110 for a NAND gate, and 1000 for a

Table 6-2

Inputs and Outputs for Two Input Gates

			Output				
Decimal number	Input 1	Input 2	AND	OR	NAND	NOR	Bit value
0	0	0	0	0	1	1	2^3
1	0	1	0	1	1	0	2^2
2	1	0	0	1	1	0	2^1
3	1	1	1	1	0	0	2^0

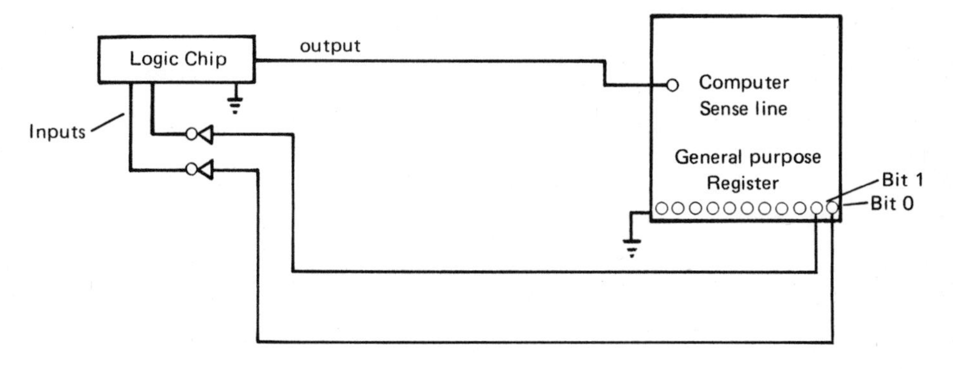

Fig. 6-1
Computer-controlled gate-test interface.

NOR gate. If these patterns are taken as binary numbers and converted to decimal, then a gate is an AND gate only if the result is decimal 1, while 7 represents an OR gate, 8 a NOR gate, and 14 a NAND gate; any other result indicates a malfunctioning gate.

The final interface configuration for a two-input gate is shown in Fig. 6-1. Note that all grounds (both on the computer interface and on the chips and chip power supply) *must* be wired together in order to complete the circuit. A logic flow chart is shown in Fig. 6-2.

EXPERIMENTAL

Equipment Needed

- 1 EL Digi-designer
- 1 Quad dual-input AND gate package (SN 7408)
- 1 Quad dual-input NAND gate package (SN 7400)
- 1 Quad dual-input OR gate package (Signetics SP 384A)
- 1 Quad dual-input NOR gate package (SN 7402)
- 1 Hex inverter package (SN 7404)
- 1 Laboratory computer system with digital I/O interface (see Appendices C and F)

Procedure

1. Connect the circuit shown in Fig. 6-1.
2. Write a program that will check any two-input gate.
3. Test several gates.
4. If time permits, modify your program to test *all* gates on the same chip, or to test gates with more than two inputs.

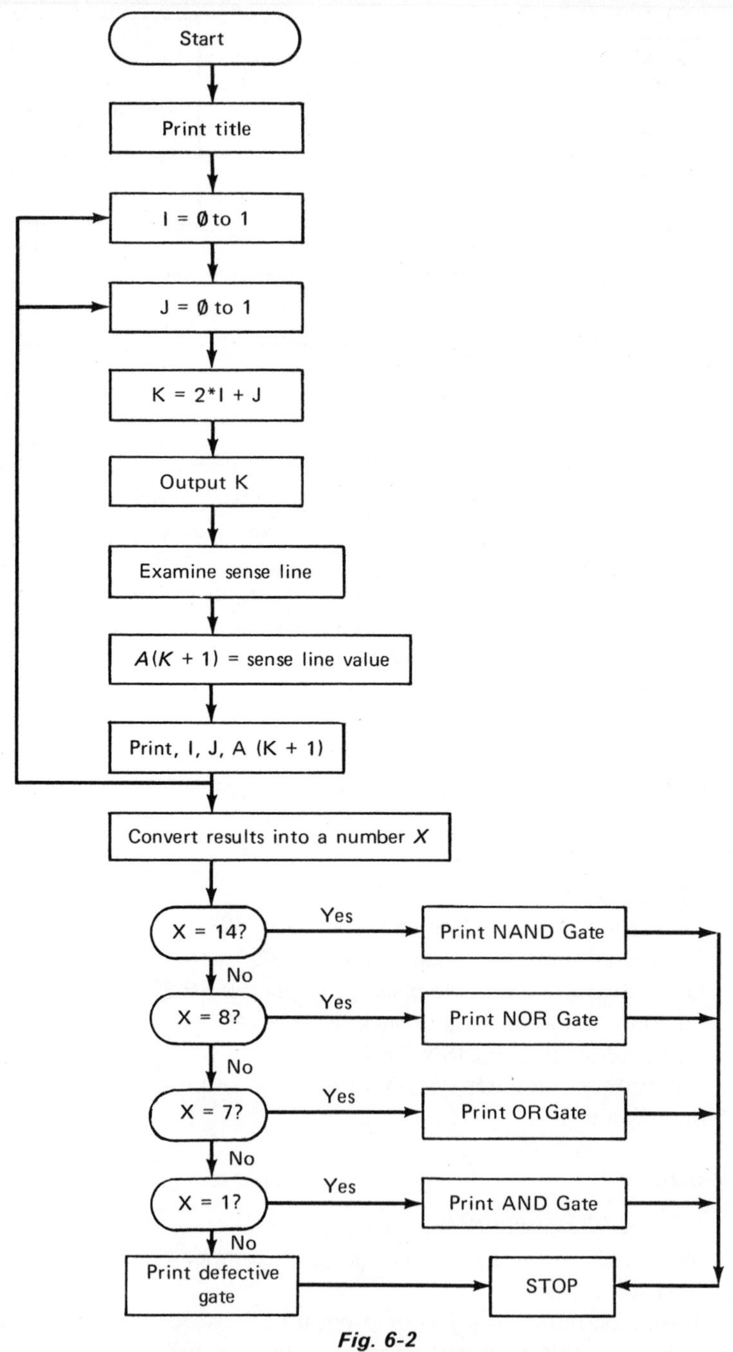

Fig. 6-2
Program flow chart for computer-controlled gate test.

Computer-Controlled Analog-to-Digital Conversion

In experiment 5, digital-to-analog and analog-to-digital conversions were discussed, and several ADC's were constructed. Although the individual characteristics of the different ADC's varied, all of them utilized the same basic design features shown in the block diagram of Fig. 7-1. Of the three functions shown, the voltage-comparison and reference-voltage-generation functions are relatively easy to implement, while the controller is considerably more difficult. A comparison of Figs. 7-1 and 5-9 shows that the first two functions can be performed by a single chip each (the 710 comparator and the Datel DAC-9), while the remaining components are all necessary for the control function. It is clear that Fig. 5-9 could be simplified considerably if one single unit could be found, which would handle all the necessary control and logic operations. It is in precisely such control and decision-making functions that laboratory computers are particularly useful.

In this experiment, the control capabilities of the laboratory computer and interface will be used with a DAC and comparator to construct several different types of ADC's. Because all three ADC's include a DAC and a

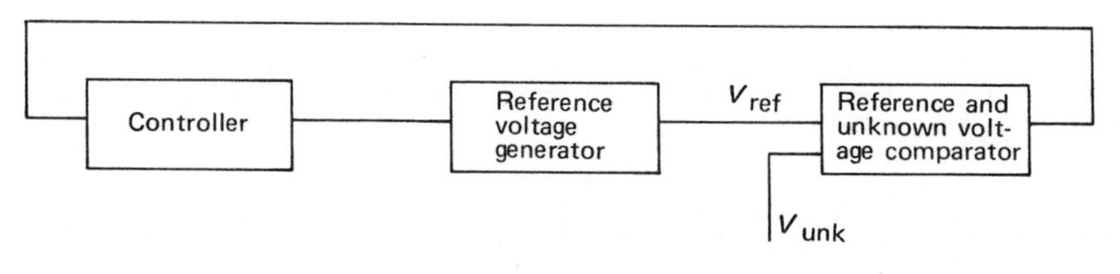

Fig. 7-1
Block diagram of an ADC.

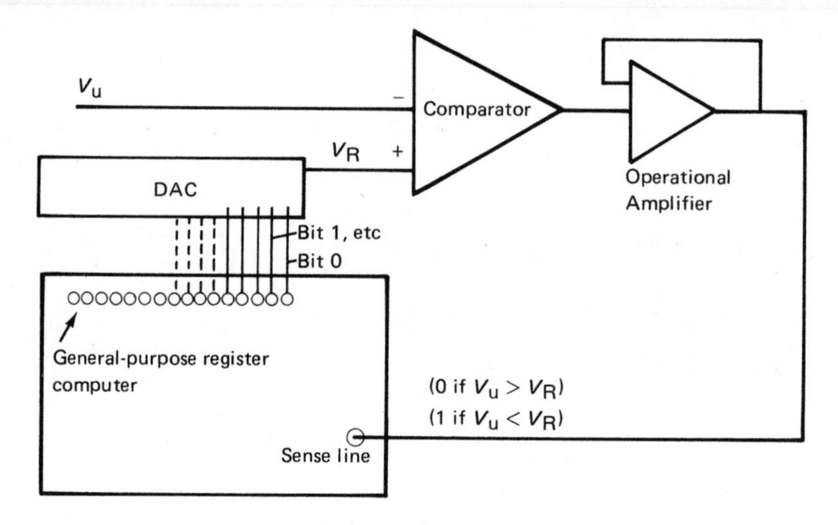

Fig. 7-2
A computer-controlled ADC.

comparator, the various techniques for conversion differ mainly in how the voltage is varied. Figure 7-2 is a block diagram of a computer-controlled ADC built in this way. By far the simplest method is for the computer to start at zero, output this value to the general-purpose register (and, therefore, to the DAC), and test the output of the comparator. The register which controls the DAC voltage is then incremented one bit at a time, until the

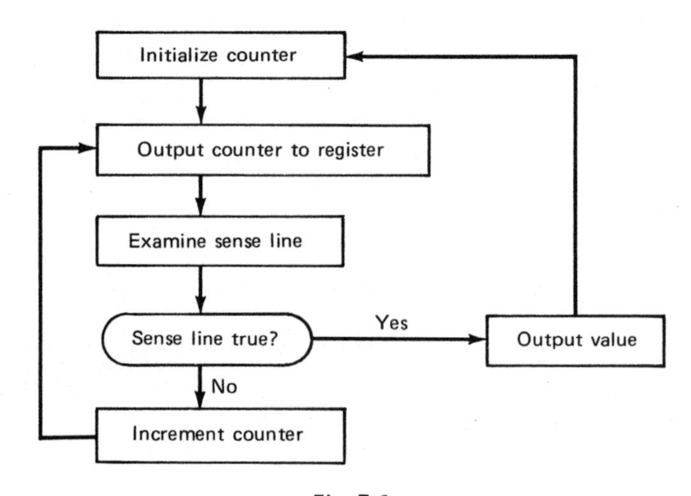

Fig. 7-3
Program flow chart for a computer-controlled staircase–ramp ADC.

DAC voltage equals or exceeds the unknown voltage (V_u). When the computer is programmed in this fashion, the ADC is of the staircase–ramp type. A flow chart for such a program is shown in Fig. 7-3.

One of the limitations of this type of computer-controlled ADC is that the computer–DAC system must be able to change the reference voltage (V_R) much faster than the unknown voltage can change for accurate results to be obtained. A minor modification of this program provides more rapid conversion rates. By starting the DAC at the last measured voltage value and counting up *or* down (depending on the initial value of the comparator), it is possible to obtain a continuous balance voltage comparison type of ADC. Figure 7-4 shows the flow chart for a program of this kind. The speed of this converter is much increased over that of the first, but still the rate of change of the input signal can be no greater than that of the least significant bit count rate.

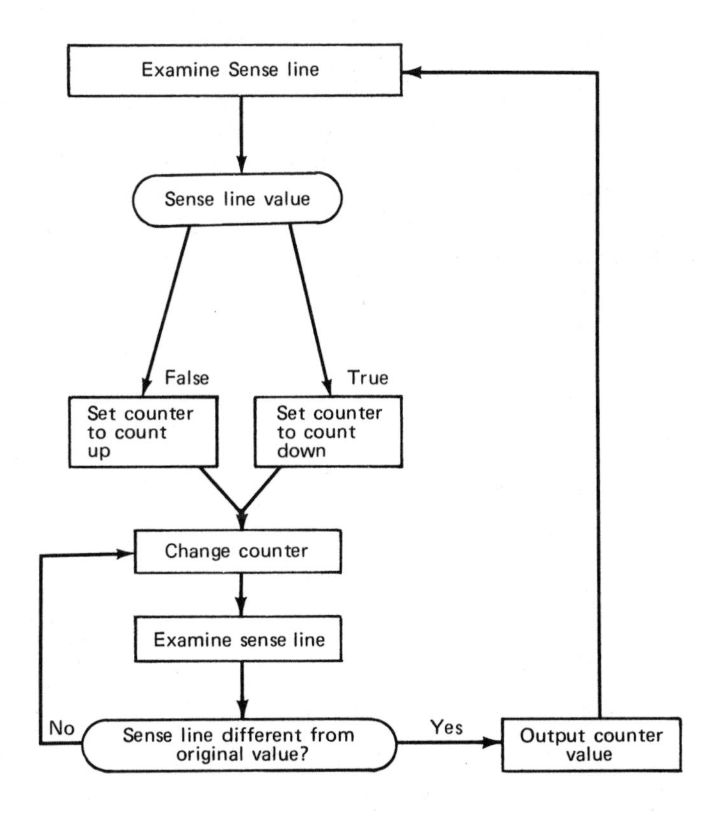

Fig. 7-4
Program flow chart for a computer-controlled continuous-balance ADC.

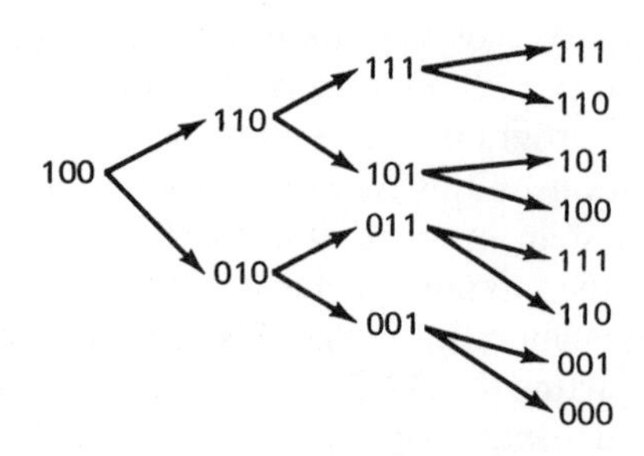

Fig. 7-5
Decision-tree for a 3-bit converter.

Successive-approximation techniques yield even faster conversion rates than the continuous converter, as Experiment 5 pointed out. Here the counter word is generated one bit at a time, starting with the most significant bit (MSB). The possible pathways for a 3-bit converter are shown in Fig. 7-5, and a typical oscilloscope display is shown in Fig. 7-6. In this type of converter, the same length of time T is required to convert each of N bits, and hence the total time necessary for a conversion is a constant ($T \cdot N$ sec). In contrast to this, the staircase–ramp converter requires $T \cdot 2^N$ sec, where 2^N is the value of the voltage to be measured. A program flow chart for a successive-approximation type ADC is shown in Fig. 7-7.

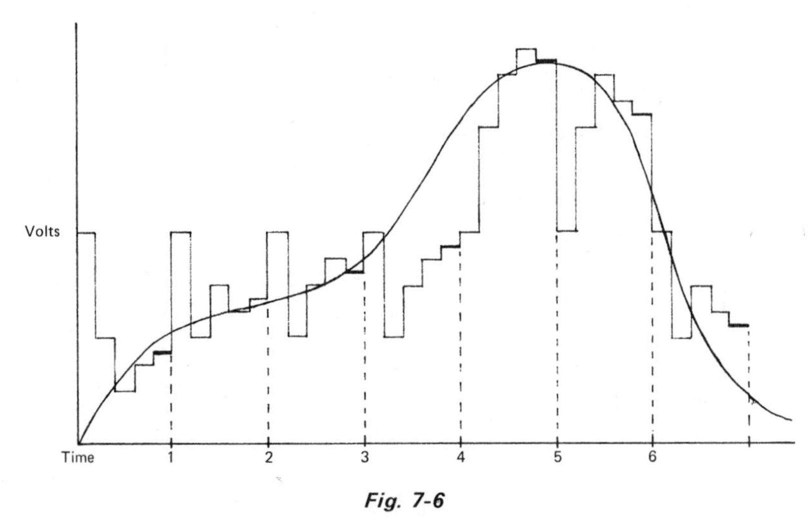

Fig. 7-6
An idealized oscilloscope display of a successive-approximation ADC in operation.

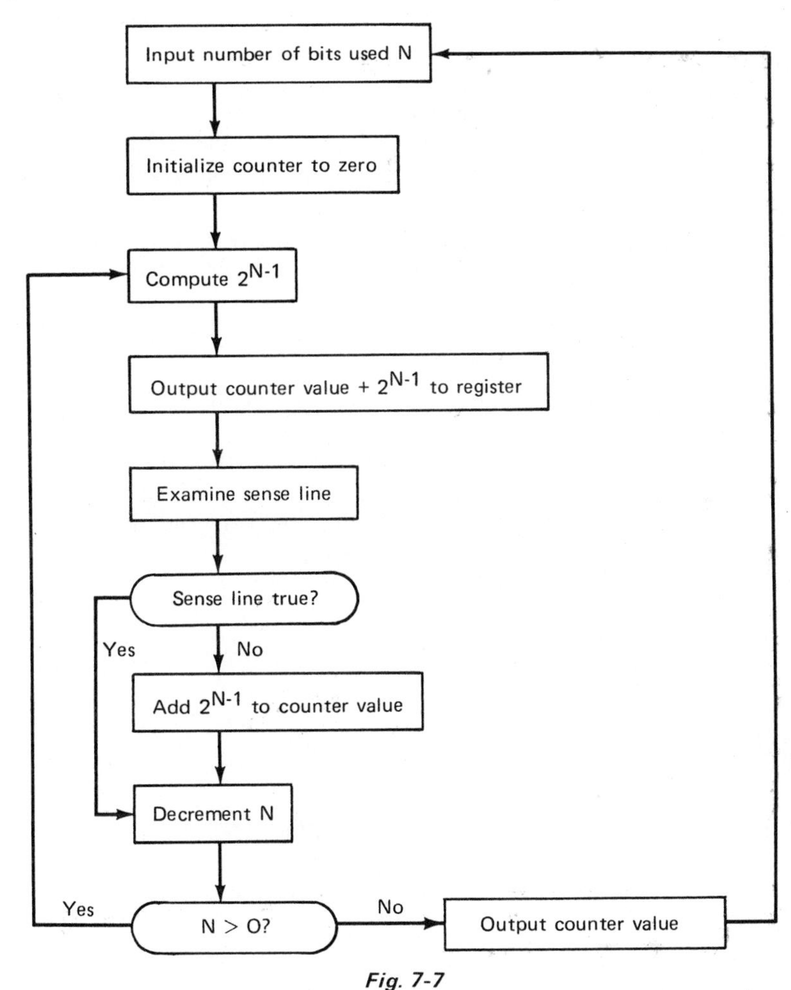

Fig. 7-7
Program flow chart for a computer-controlled successive-approximation ADC.

EXPERIMENTAL

Equipment Needed

- 1 EL Digi-designer
- 1 ±15-V power supply
- 2 Variable power supplies (0 to 15 V)
- 1 Voltage comparator (LM 710C)
- 1 Operational amplifier (SN 72741)
- 1 Digital-to-analog converter (Datel DAC-9)
- 1 Laboratory computer system with digital I/O interface

Procedure

1. Connect the circuit shown in Fig. 7-2. If the computer interface contains a DAC, this may be used in place of the register and DAC combination shown. Use a minimum of four bits on the DAC. On some computer systems, a voltage follower is needed between the comparator output and the sense line input (the operational amplifier just to the right of the comparator). Your instructor will tell you whether this is required.
2. Write and execute a program to find an unknown voltage by the staircase–ramp method.
3. Write and execute a program to find an unknown voltage by the continuous-balance method.
4. Write and execute a program to find an unknown voltage by the successive-approximation method.
5. If a computer-readable clock or timer is available, determine the length of time required for each conversion.

Experiment 8

Graphic-Display Experiments Using an X–Y Plotter and/or Storage Oscilloscope

The computer provides a speedy method for the evaluation and display of functions. This is highly desirable in curve fitting, data smoothing, or plotting of real or simulated data. In order that he may use these capabilities, it is necessary that the user understand the particular problems associated with plotting functions. It is the objective of this experiment to provide an opportunity for the student to develop an appreciation of the use of graphic-display techniques through programming exercises involving both simple function plotting and data-smoothing methods.

A laboratory computer can output voltages *via* a digital-to-analog converter (DAC) to external devices. If these happen to be display devices, such as an X–Y plotter or an oscilloscope, pairs of such values may define points to be plotted or displayed and a series of such points—a plot. Plots may either be generated from data previously acquired from an instrument, digitized, and stored in memory, or from points resulting from the evaluation of mathematical functions. The same basic problems exist whether experimental data or calculated values are plotted.

It is of fundamental importance to scale the data appropriately so that it will fit into the space available. Therefore, care should be taken to check the voltage-input range of the plotter or oscilloscope in order to assure that the output of the DAC falls within the limits. Figure 8-1 contains a general flow chart for programs which evaluate functions and plot them using an X–Y plotter. Figure 8-2 is a sample REAL-TIME BASIC program which, on execution, will generate the "3-leaved rose" shown in Fig. 8-3. Notice that in the sample program the origin is first set, then the value of the function is appropriately scaled. Subroutine "ODAC" outputs the first argument to the X-DAC, and the second argument to the Y-DAC.

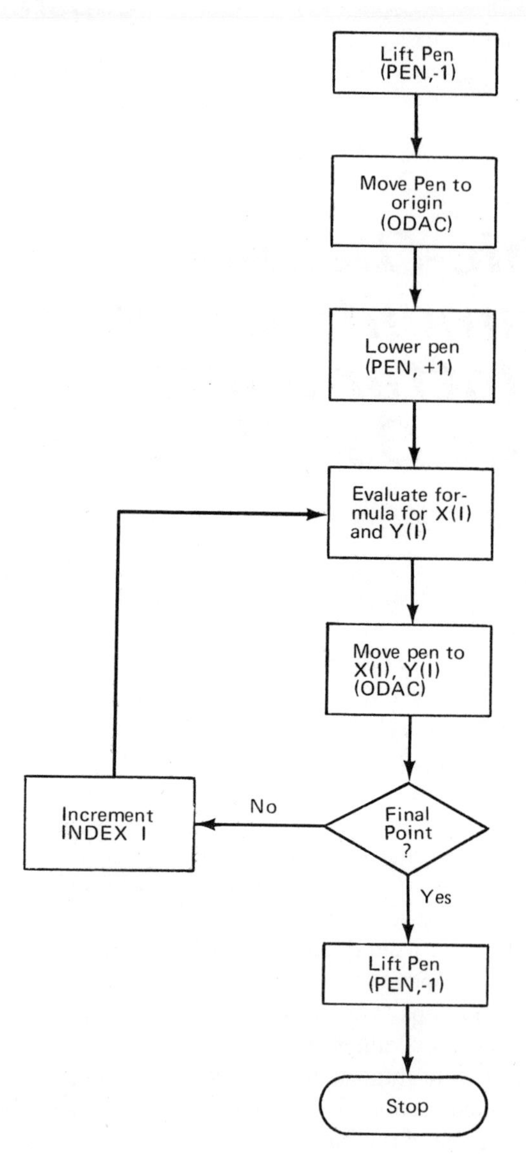

Fig. 8-1
A general flow chart for programs to evaluate functions and plot them on an X–Y recorder.

One of the important aspects of computer-controlled plotting of experimental data is the flexibility of the preprocessing methods which may be applied. One of the most common ways to treat such data is the application of mathematical-smoothing techniques in order to decrease noise (or, to

```
10    REM 3-LEAVED ROSE
20    CALL PEN, -1
30    LET A = -3.14159
40    LET R = 10*COS(3*A)
50    LET X = R*COS(A)
60    LET Y = R*SIN(A)
70    CALL ODAC(X, Y)
80    CALL PEN, +1
90    FOR A = -3.14159 TO 3.14159 STEP 0.005
100   LET R = 10*COS(3*A)
110   LET X = R*COS(A)
120   LET Y = R*SIN(A)
130   CALL ODAC(X, Y)
140   NEXT A
150   CALL PEN, -1
160   END
```

Fig. 8-2

A REAL-TIME BASIC program to plot a 3-leaved rose.

Fig. 8-3

A 3-leaved rose.

put it another way, to improve the *signal-to-noise ratio*). The procedure we will briefly discuss here is that of Savitzky and Golay (see refs. D-9 and D-10). In data smoothing, a number of restrictions must be observed. First, the data points must be located at fixed and uniform intervals along the abscissa. Second, the data curve must be a continuous function. The simplest method of data smoothing is called the "moving-average" method. In this procedure, a fixed odd number of successive data points are averaged. The center point of the group is assigned the average value. Next, the first data point is dropped from the group, and the next unused data point of the total set is incorporated into a new average. This process is repeated until all remaining data points are treated. Note that the procedure results in the loss of $(N - 1)/2$ data points at the beginning and at the end of the data stream for each smooth, where N represents the number of points used in the average. This method assigns equal weight to all the data points.

A modification of the technique is to weight each of the data points used in the "average" differently. This, of necessity, will allow some points to exert a greater influence on the value of the smoothed point than others. The weighting factors are called convolution integers. Figure 8-4 shows several possible weighting functions over the data interval used. These schemes are useful in reducing the noise, but will degrade data points which have real, sharply different values.

Recall that there is a mathematical procedure, the method of least squares, which will allow determination of the best equation to fit a set of

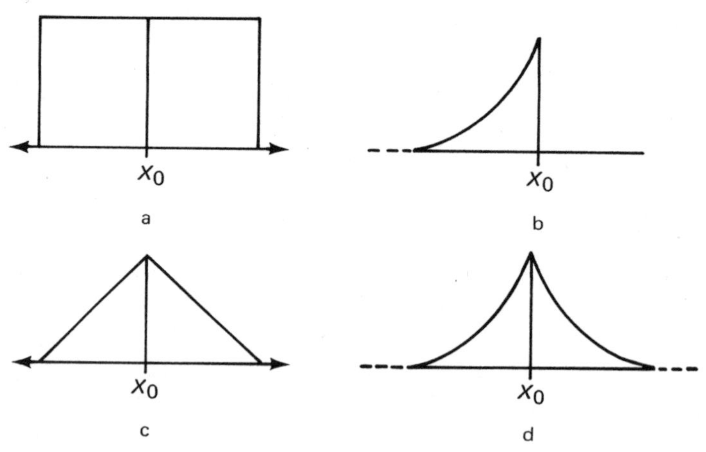

Fig. 8-4

Various convolute functions. (a) Moving average; (b) exponential function; (c) symmetrical triangular function, representing idealized spectrometer slit function; (d) symmetrical exponential function. [Reprinted with permission from A. Savitzky and M. J. E. Golay, *Anal. Chem.*, **36**, 1627 (1964). Copyright by the American Chemical Society.]

values. If one chooses a small contiguous portion of the larger set of data points, then an equation which best describes these data can be determined. If, then, one calculates the ordinate value at the center of these contiguous data points, it will be the best possible value. Further consideration of the least-squares method and the fact that we are interested in only one ordinate value leads to the realization that the coefficients in the equation derived are needed only to evaluate the one central point. It is therefore possible to derive a set of convolution integers, the values of which will depend on the equation used and the number of points used in the smoothing step. Savitzky and Golay have evaluated these integers for a variety of equations and different numbers of data points used in the smooth. Table 8-1 is reproduced from reference D-9 and contains the convolution integers for either quadratic or cubic equations for any number of points from 5 to 25.

Table 8-1*

Convolution Integers for Either Quadratic or Cubic Smoothing Functions

Points	25	23	21	19	17	15	13	11	9	7	5
−12	−253										
−11	−138	−42									
−11	−33	−21	−171								
−09	62	−2	−76	−136							
−08	147	15	9	−51	−21						
−07	222	30	84	24	−6	−78					
−06	287	43	149	89	7	−13	−11				
−05	343	54	204	144	18	42	0	−36			
−04	387	63	249	189	27	87	9	9	−21		
−03	422	70	284	224	34	122	16	44	14	−2	
−02	447	75	309	249	39	147	21	69	39	3	−3
−01	462	78	324	264	42	162	24	84	54	6	12
00	467	79	329	269	43	167	25	89	59	7	17
01	462	78	324	264	42	162	24	84	54	6	12
02	447	75	309	249	39	147	21	69	39	3	−3
03	442	70	284	224	34	122	16	44	14	−2	
04	387	63	249	189	27	87	9	9	−21		
05	343	54	204	144	18	42	0	−36			
06	287	43	149	89	7	−13	−11				
07	222	30	84	24	−6	−78					
08	147	15	9	−51	−21						
09	62	2	−76	−136							
10	−33	−21	−171								
11	−138	−42									
12	−253										
Norm	5175	805	3059	2261	323	1105	143	429	231	21	35

*Reprinted with permission from A. Savitzky and M. J. E. Golay, *Anal. Chem.*, **36**, 1627 (1964). Copyright by the American Chemical Society.

The signal-to-noise ratio is increased by roughly a factor of the square root of the number of points used in each "average," i.e., a 9-point smooth gives thrice the original signal-to-noise ratio, a 5-point about twice, etc. If the data density is moderate to high, then most peaks of experimental curves, such as spectra, can be described by a quadratic equation, and the shoulders by cubic equations. The information in Table 8-1 is adequate for this purpose. The procedure used is the same as that first described for the "moving average," except that the weighting integers and the divisors, or normalizing factors, are given in the table. Thus, to smooth point 5, using a nine-point quadratic or cubic data smooth, point 1 is multiplied by -21, 2 by 14, 3 by 39, 4 by 54, 5 by 59, 6 by 54, etc., and the products are summed and divided by the normalizing factor 231. For smoothing point 6, point 1 is dropped and point 10 brought in and the procedure repeated. Savitsky and Golay also give tables for derivative functions (1st, 2nd, and 3rd) and for polynomials through degree 5. Remember that derivative functions have usefulness in distinguishing changes in curve shapes.

This experiment is designed to give practice in using oscilloscope and plotter routines and improving signal-to-noise ratios by application of the smoothing techniques of Savitzky and Golay.

EXPERIMENTAL

Equipment Needed

1 Laboratory computer system with X–Y plotter and/or storage display oscilloscope interface

Procedure

A. Plotting

Write a program to plot on the oscilloscope or plotter each of the following figures.*

1. Straight line ($y = mx + b$)
2. Astroid ($x = a \cos^3 \theta$, $y = a \sin^3 \theta$)
3. N-leaved rose (N may equal 1, 2, 3, 4, or 5) ($r = a \sin N\theta$)
4. Spiral of Archimedes ($r = a\theta$)
5. Exponential ($y = a^x$)

*To transform polar to Cartesian coordinates, remember $x = r \cos \theta$, and $y = r \sin \theta$.

B. Data Smoothing

1. Write a program to evaluate and print the function $y = x^3$ at 100 equally spaced values of the abscissa.
2. Add to the program a section to allow you to arbitrarily add some "noise" to several of the y values, noting which they are.
3. Write a program to determine the moving average, convolution integers all equal to 1, for a 9-point smooth of the 100 data points. Run the program.
4. Change the program to use a 9-point smooth with the convolution integers shown in Table 8-1. Run this program. Compare the results with those of steps 1, 2, and 3.

Experiment 9

Transient Decay Signal (Capacitor Discharge)

Experiments 3, 4, and 5 of the Logic Experiment section introduced you to the various devices which are necessary in order for the computer to acquire analog data in useful or digitized form. It is, of course, necessary that these devices be combined in such a way as to allow the computer complete control of the direct entry of the data. The necessary devices generally consist of four parts: an analog-to-digital converter to digitize the analog signal, a buffer register to store the digitized data temporarily, a time-base generator to determine the appropriate digitization frequency, and, finally, control logic to properly sequence the digitization of the analog signal and the transfer of the digital value into the computer.

The time-base generator or clock is usually a high-frequency crystal-based oscillator. There are in common usage crystals with frequencies up to about 20 MHz. Longer time bases than this are obtained by using an appropriate counting circuit so that only every Nth (e.g., 100, 1000, or even 10^6) crystal oscillation is output by the timing circuit. The time base or frequency should be selected according to the rules discussed in Experiment 12. For this experiment, however, simply use the data rates suggested.

This data-acquisition frequency may be quite low, as in the case of a very slowly changing gas chromatograph signal, or it might be quite high, as is often the case in mass-spectrometry applications. The proper time-base selection obviously requires knowledge of the rates of change expected in the analog signal. Experiment 12 discusses in more detail the problems associated with selection of appropriate sampling frequencies. It should be remembered that the faster the conversion rate, the shorter is the total elapsed time over which the experiment can be examined, since the computer's memory will have a finite (often small) space available for storage of data. It becomes necessary, then, to make every effort to use that space as efficiently as possible. It is possible that signals from some experiments may be sampled at different frequencies during the lifetime of an experiment. For example, in kinetics work the rate of change of concentration of a reactant species will change with time (if the order is nonzero). By adjusting the frequency of data acquisition, we can condense data from a large time domain into the limited space available in the computer. So long as the data taken adequately represent the experiment, we should be able to construct the data curve from the small number of points with the same confidence as if we had a large number of data points.

Figure 9-1 illustrates an analog signal which can be condensed in a data buffer by acquiring data from it at different rates. It should be obvious that in order to adequately define the curve, data will have to be taken at closely spaced intervals in the region marked A, but that in the region labeled B much larger sampling intervals can be used, and good definition of the curve will still be obtained. Any data signal which has regions of substantially different rates of change can be followed adequately using a variable data-acquisition rate.

There are at least two alternatives in taking data at different rates. The easiest is to set in the program the number of data points to be taken at each frequency. This requires some foreknowledge as to the exact nature of the waveform to be sampled. A second and more sophisticated approach is to allow the computer to calculate the rate of change and adjust the data-acquisition rate accordingly. This latter approach, however, has the disadvantage that if very rapid data rates are required, the computer may not have sufficient time to perform the necessary calculations required to correctly readjust the data-acquisition rate. Thus, one might be forced to set the upper limit of data-acquisition rate to a value much slower than the absolute maximum allowed by the computer system.

The higher-level languages such as BASIC or FORTRAN are inherently slower than those cases where the program statements were written step by step in the assembly language of the particular computer. (However, FORTRAN will usually be faster than BASIC, since BASIC recompiles each statement as it executes it.) For example, a statement in BASIC to add A to B might require execution of 400 assembly-language instructions, whereas the same function perhaps could have been done in 30 assembly-language instructions if written as such. Since each instruction requires a known and finite amount of time to execute, a reduction of the number of instructions cuts the time required for any calculation. Should the data-acquisition-related program segments be written in the assembly language, it is possible to save relatively large amounts of time, thereby increasing the maximum rate at which data can be acquired.

In this experiment, you will monitor an exponential decay waveform as shown in Fig. 9-1. The waveform is generated by the discharge of a charged capacitor through a resistor. The discharge can be triggered by an external electronic signal provided by the computer. See Fig. 9-2 for a schematic of

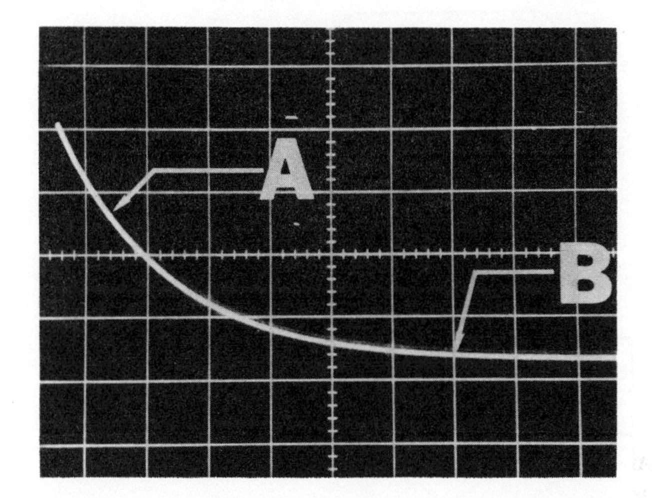

Fig. 9-1
A capacitive-discharge decay curve.

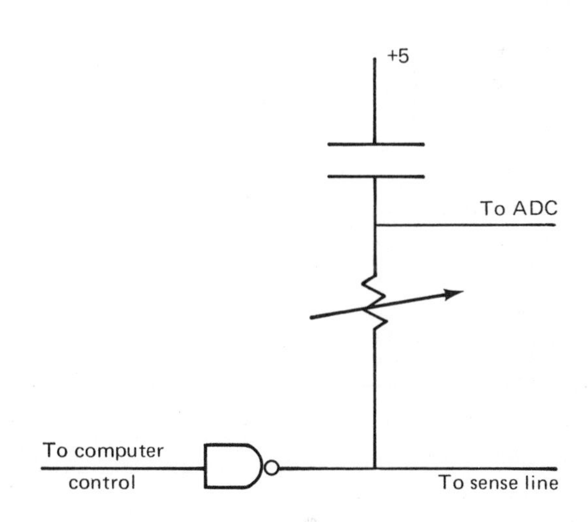

Fig. 9-2
A computer-controlled capacitor discharge circuit.

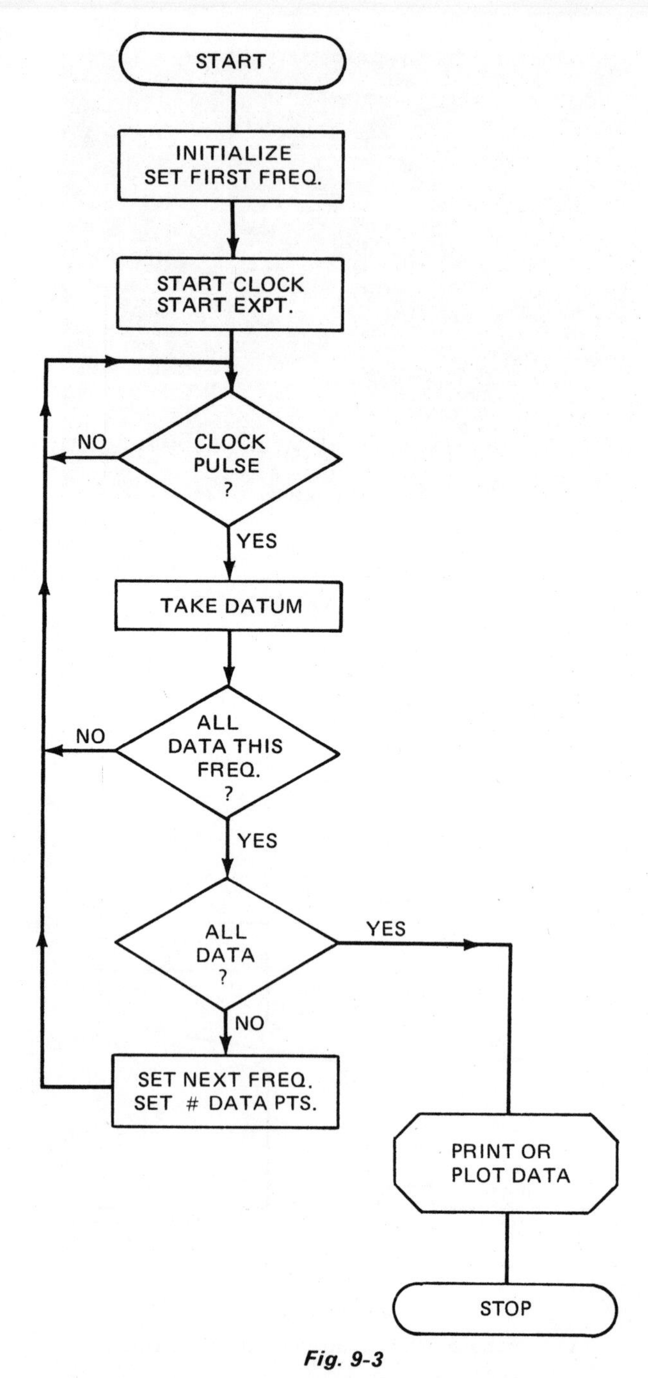

Fig. 9-3
Flow chart of a variable-rate data-acquisition program.

the simple circuit required. (A more complex circuit, which superimposes noise on the signal, is given in Appendix E.) Figure 9-3 shows the flow chart for a program which includes a variable data-acquisition rate. This experiment is designed to teach the principles of adjusting data-acquisition rates to be compatible with the rate of change of the data being acquired.

EXPERIMENTAL

Equipment Needed

1 EL Digi-designer
1 Hex inverter package (SN 7404)
1 Resistance substitution box
1 Capacitance substitution box
1 Laboratory computer system with programmable clock and digital and analog I/O interfaces

Procedure

1. Connect the computer, general-purpose interface, and circuit as shown in Fig. 9-2. Do not use a resistor value greater than 100 kΩ, or the inverter may not operate properly. Remember to connect all the grounds together, even though this is not shown in Fig. 9-2.
2. Write a program to acquire data, as the capacitor discharges, *either* at three or four different rates (i.e., 10 points at 200 Hz, 10 points at 20 Hz, 10 points at 2 Hz), *or* to continuously increase the time between data-point acquisition. (The latter approach gives better results but precludes initial acquisition rates much faster than 50 Hz.)
3. Run the program in step 2. Now modify the program so that the experiment initiation step is under computer control. Run this program.
4. Is the acquisition scheme you are using adequate to give a good description of the curve? If not, revise the program to give a more optimum sequence, and rerun the program.
5. Write a program to produce a semilog plot of the data and to find the linear least-squares best fit of the logarithmic data. Notice that this capacitive discharge is a logarithmic decay analogous to a first-order chemical reaction. Keeping this in mind, what is the significance of a semilog plot of the data?

Experiment 10

Ensemble Averaging of Repeatable Noisy Signals

All signals produced by instruments contain various systematic and nonsystematic components which are not produced by the phenomenon of interest (e.g., random noise generated by electronic components, background absorption in spectrophotometers, etc.). Collectively, all components which are extraneous and which the scientist would prefer to have entirely eliminated from the signal are called "noise." Clearly, noise may be either random or systematic. Various methods may be employed to minimize or eliminate noise, but all depend upon a detailed understanding of the source of the noise and its relationship to the signal of interest. In many cases, careful hardware design can minimize noise via judicious use of electronic filters, good wiring practices, and the like. It is, however, often the case that due to experimental limitations it is necessary to accept a low signal-to-noise ratio from an instrument. In this situation, provided the noise has certain characteristics, it may become possible to significantly improve the signal-to-noise ratio by using a combination of appropriate data-sampling techniques and mathematical analysis. In this experiment, the use of such methods to eliminate *random noise* will be explored.

The two conditions which the noise must meet if it is to be amenable to the application of the signal-enhancement method, which is the subject of this experiment, are simple ones. First, the noise must be random, or if not random, at least uncorrelated with the signal of interest. Second, the experiment (with the noise superimposed on it) must be reproducible. Ideally, noise arises from the sum of a very large number of totally uncorrelated events, such as thermal motion of nonrelated atoms in a conductor; therefore the probability function for the noise voltage is purely Gaussian, as shown in Fig. 10-1, and has the form Ke^{-V^2/W^2}, where K is a constant, V is the voltage and W is the width of the probability function which determines the actual noise levels. To determine the probability of two events in succession, one must take the product of the probabilities of the events occurring individually. Therefore, the normalized probability function for two repeated experiments

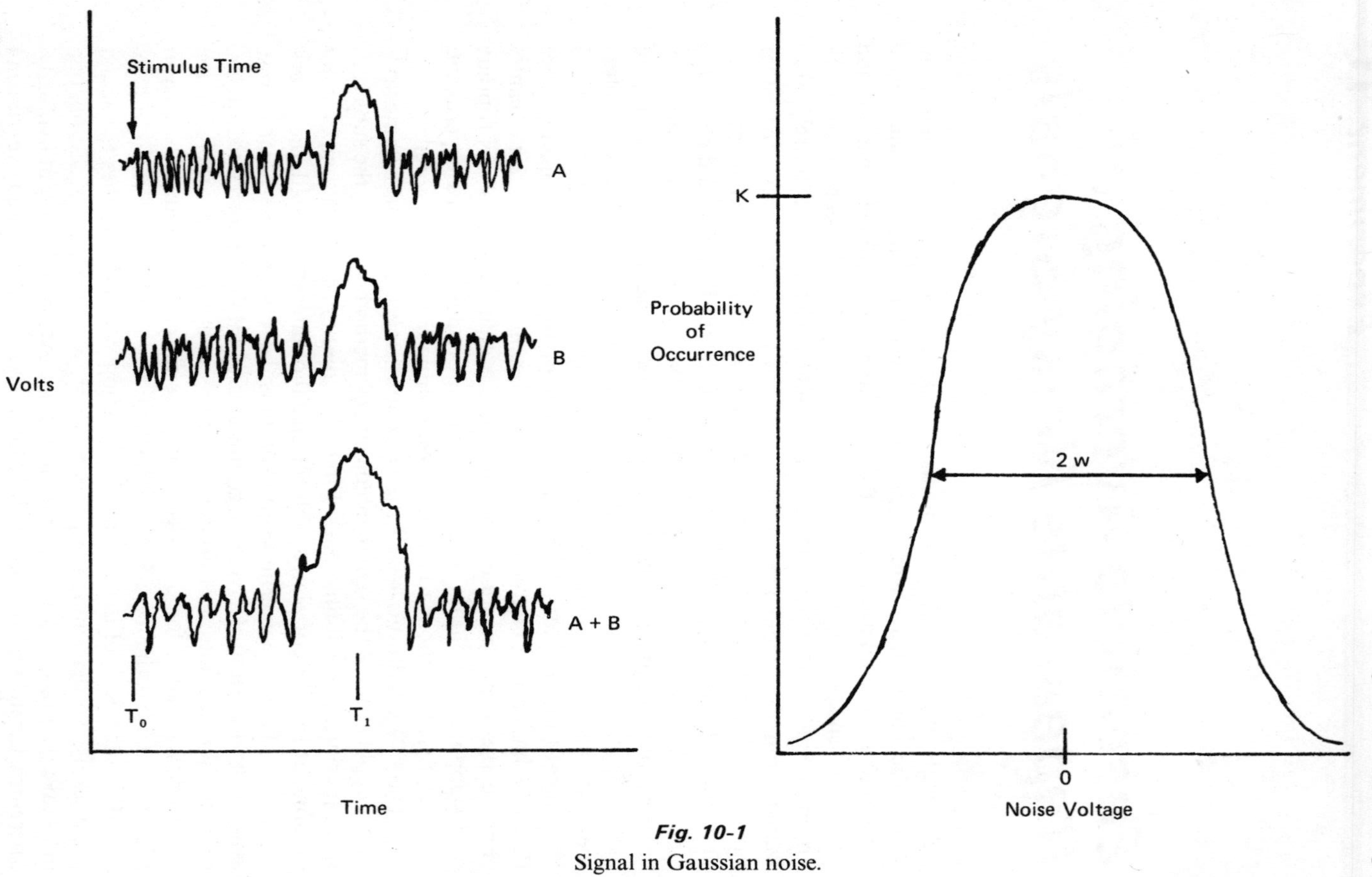

Fig. 10-1
Signal in Gaussian noise.

is Ke^{-2V^2/W^2}, and for N experiments it is Ke^{-NV^2/W^2}. To determine the noise level for N combined experiments as compared to that of a single experiment, merely solve Eq. (10-1). This predicts an increase of a factor of $\sqrt{N}$.

$$Ke^{-V^2/W_s^2} = Ke^{-NV^2/W_N^2} \qquad (10\text{-}1)$$

$$W_N = \sqrt{N}W_s$$

The signal itself grows linearly with N, since it is assumed with a probability of 1 that the response to the stimulus will occur at time T_1. Therefore, the signal-to-noise ratio increases according to Eq. (10-2).

$$(\text{Signal/noise})_N = S_s N/\sqrt{N} = S_S\sqrt{N} \qquad (10\text{-}2)$$

This method of signal enhancement is called "ensemble averaging." In order to normalize the summed signals, the totals are divided by the number of individual experiments utilized.

Figure 10-2 shows a triangular signal with superimposed random noise. Figure 10-3 shows the same signal after averaging of 100 repetitive runs. Theoretically, the signal-to-noise ratio increases as the square root of the number of experiments averaged. Thus, a doubling of the signal-to-noise ratio requires four runs averaged, whereas a 10-fold increase requires 100 runs, etc.

One serious source of error when applying ensemble averaging is related to the failure to initiate the addition sequence at exactly the same point in the signal life. This will cause the average to be biased and distorted. The result is that, although the noise still changes as the $\sqrt{N}$, the average does

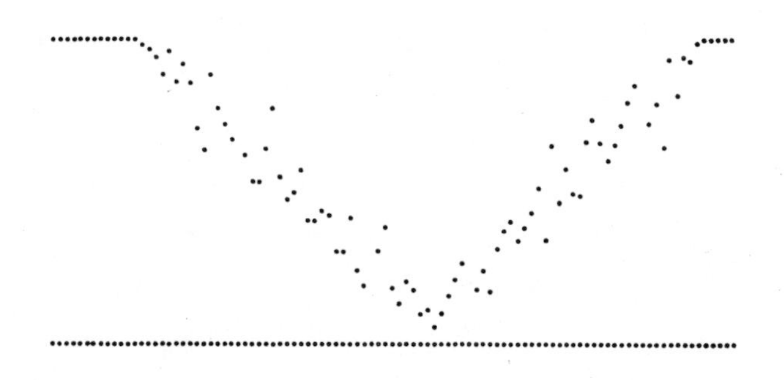

Fig. 10-2
A triangular wave with superimposed noise.

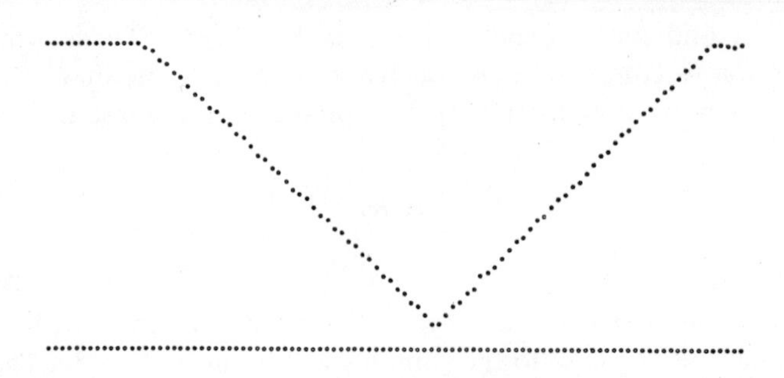

Fig. 10-3
The triangular signal after ensemble-averaging 100 experiments.

not represent the true signal due to time drift in T_1. Figure 10-4 shows the effect of averaging the same repeatable triangular signal of Fig. 10-2, except that the start of the averaging process did not always occur at the same point in the signal lifetime. This can happen when the start of the clock controlling the sampling and the start of the signal are not synchronized, i.e., when the clock is not restarted at the start of each signal to be added. Notice that the average has become broader and less distinct. It may often be necessary to tolerate some minimum delay time until the signal can be repeated. For example, if one is averaging a nuclear-magnetic-resonance signal, it takes a finite and not insignificant amount of time to reset the sweep at its starting point.

The technique of ensemble averaging is used extensively for signal-to-noise improvement in time-domain experiments where no other option

Fig. 10-4
Ensemble average of the triangular signal with synchronization error.

exists since the rate of appearance of the signal is uncontrollable. Likewise, in frequency-domain experiments, ensemble averaging is used to eliminate the effects of slow time-domain drifts, as occur in most magnetic resonance experiments.

A laboratory computer with general-purpose interface provides a very convenient way of carrying out an ensemble-averaging experiment. Such a system can be programmed to carry out a specific sequence of events, store data, carry out mathematical operations on the data, and finally output the data in usable form. The computer also can perform special types of calculations (such as digital smoothing) upon the data to improve its characteristics even more (see Experiment 8). These are generally not done in real time but rather after all the data have been collected.

The program which one uses to acquire the data for ensemble averaging must have certain characteristics to be successful. The program must include the following:

1. Selection of a clock frequency
2. Number of data points to be collected per run
3. Number of runs
4. Allocation and clearing of data-storage locations
5. Initiation of a source signal simultaneous with data acquisition
6. Summation of the signal with associated noise point by point
7. Termination of data acquisition and stopping of clock
8. Averaging routine
9. Display of ensemble average

Figure 10-5 shows a flow chart of a typical program for carrying out an ensemble-averaging experiment. If the rate of data acquisition is very rapid, the program should be written so that the data for each run is first acquired, then added to the data for the previous runs. It is also helpful if the program provides for additional runs to be made and added to previously collected data rather than requiring each new ensemble-averaging test to begin with the first run.

The purpose of this experiment is to familiarize you with the process of ensemble averaging, setting up the hardware for synchronization of data acquisition with a repetitive signal, and output of the reduced data in usable form.

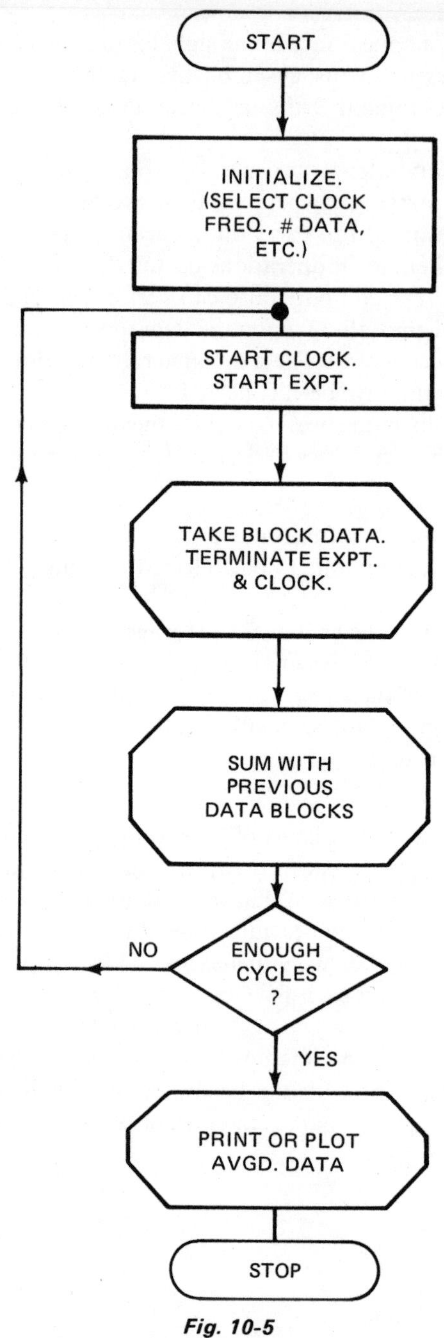

Fig. 10-5

Flow chart for ensemble-averaging program.

EXPERIMENTAL

Equipment Needed

1 Laboratory computer system with programmable clock and digital and analog I/O interfaces
1 Noisy triangular wave generator (see Appendix E for detailed description of such a generator)
1 EL Digi-designer
1 Oscilloscope or plotter (could be Teletype) for output of data waveform

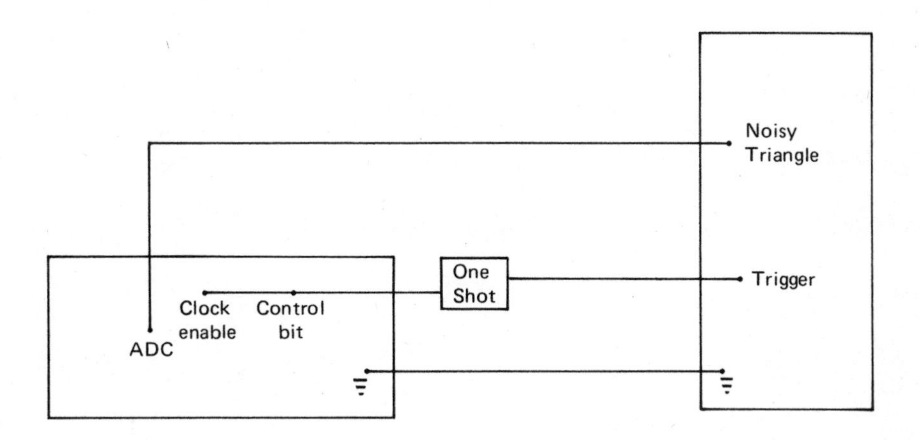

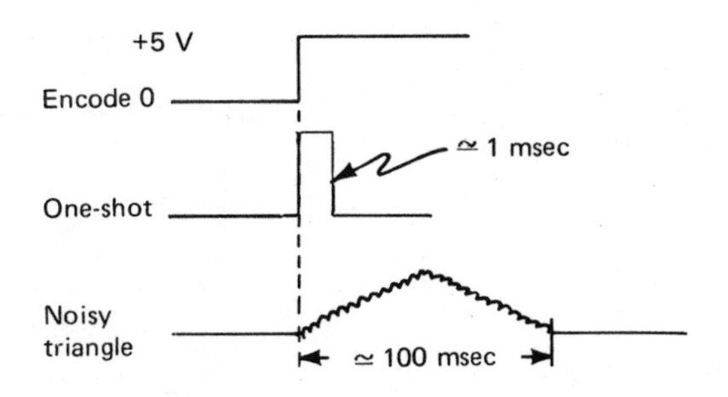

Fig. 10-6

Wiring diagram for multifunction generator-interface. Generator and converter triggered simultaneously.

Procedure

1. Set up the equipment as shown in Fig. 10-6. Be sure to include a provision for *resetting* the control bit (not shown in Fig. 10-6).
2. Write a program corresponding to the flow chart of Fig. 10-5. Include programming to plot the data on the Teletype or an interfaced plotter or oscilloscope.
3. Take data for 1, 10, 50, and 100 runs. Output the averaged data after each group of runs.
4. Visually compare the averaged data obtained in step 3 for the various runs. Does the signal-to-noise ratio appear to depend inversely on the square root of the number of averaging cycles?

Experiment 11

Multiplexing Using Triggerable Distinct Waveforms

It is very often desirable to use a single transmission channel or wire for the sequential transmission of signals derived from several sources. This method can be thought of as a simple "time-sharing" approach to the use of the channel, in that each signal source is allotted a certain portion of the available time for its use of the channel. Obviously, if each of N signal sources is allocated an equal portion of the total time available, then $1/N$ is the fraction available for any single source. This procedure is called "multiplexing" and may be used equally well with either digital or analog signals. These two types of multiplexing are represented schematically in Fig. 11-1.

In the present experiment, we will consider the question of analog multiplexing since this is a method which permits the laboratory-computer user to share a single analog-to-digital converter (ADC) among several signal sources. It is evident that such an approach, which eliminates the need for several additional ADC's, can have a great deal of practical importance in common laboratory applications. The required sequential connection of each of the signals can be accomplished by attaching each signal via a separate line/switch combination to the input channel (as depicted schematically in Fig. 11-1a). The switching could be done mechanically using a motor-driven rotary switch. However, at the speeds normally required by laboratory applications and the computer, electronic switches are usually necessary. So that no information is lost, the rate of switching from signal to signal must be much greater than the rate of change of any particular signal. Such a multiplexer is a set of electronically controlled switches which can operate at reasonably high frequency. One switch is required for each signal.

Either relays or solid-state devices, usually field-effect transistors (FET) which can be turned on or off depending upon the voltage level applied at a "gate" terminal, are used (see Experiment 4). As long as the gate is "on," a signal can traverse the switch unimpeded. If the gate is "off," transmission of

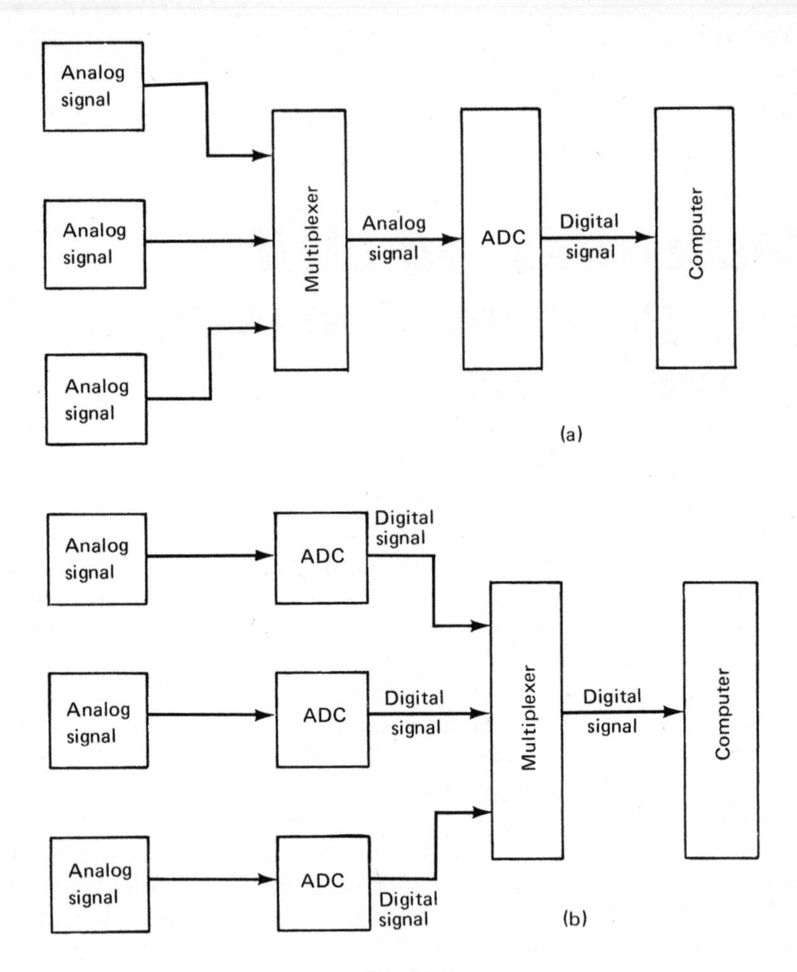

Fig. 11-1

Block diagrams of multiplexing arrangements. (a) Analog multiplexing; (b) digital multiplexing.

the signal is inhibited. Since FET devices are relatively easy to implement in this way, one option in the present experiment is the construction of a simple multiplexer from such components. Before discussing that, however, let us briefly return to some more general considerations.

The multiplexer, as we have defined it, provides for no control of switching rate or sequence. Both the sequence and frequency must be controlled by additional logic circuitry. In usual applications, the analog multiplexer output is directly connected to the input of an ADC. So common is this particular arrangement that many commercially available ADC's have multiplexers provided with them as either standard or optional features. One problem which can occur is caused by a change in the signal which takes place while it is connected via the multiplexer to the ADC. If this happens, the converted signal value will not be accurate. In order to overcome this problem, track (or sample)-and-hold amplifiers are used (see Experiment 4). Such devices follow the signal level at the output of the multiplexer. When the multiplexer switches input channels, the T/H amplifier can be gated to "hold" the signal level during digitization. The value obtained therefore corresponds to the signal value at the instant the gating takes place. After digitization is complete, the amplifier can be gated back to the "track" mode. For the reasons we have just discussed, T/H amplifiers are a very common requirement for multiplexer–ADC combinations and are also generally available as either standard or optional parts of commercial ADC packages.

When only two lines are to be sampled alternately, construction of a multiplexer is relatively simple. Figure 11-2 shows a simple diagram for multiplexing two lines using two track-and-hold amplifiers and two multiplexer switches. This operates simply as follows: When the clock pulse is low, FET-1 is on, allowing the T/H to follow input signal 1. During this time, FET-2 (to the ADC) is off, FET-3 is off (i.e., the T/H amplifier for signal 2 is in the "hold" mode), and FET-4 connects the held signal to the ADC. When the clock goes high or true, the reverse situation occurs and signal 1 reaches the ADC through its T/H amplifier. Thus, the two channels are sampled alternately at the rate of the clock frequency. When more than two lines (channels) are to be sampled, or when the rates of sampling the channels are different, additional logic circuitry must be used. One approach is to use simple counters driven by a clock to control the timing. These systems can be designed to allow for different rates of sampling. Alternately, if the same rate of sampling is to be used, then simple gates or shift registers can be used for control. Another method is to gate the various switches in such a way that they are "on" only when a certain bit pattern is output from the computer. The computer, therefore, controls which channel is connected to the ADC. This is the technique you will use in the first part of this experiment. There is also an implicit assumption that the actual conversion time of the ADC is small compared to the switching frequency.

The multiplexer shown previously in Fig. 11-2 does not allow both T/H amplifiers to be held at the same instant in time. The circuit could be changed in such a way that both could be held with the same clock pulse. The sequence of service to the various channels could be controlled separately so that each channel is consecutively serviced under program control.

In this experiment, you will become familiar with the use of multiplexing

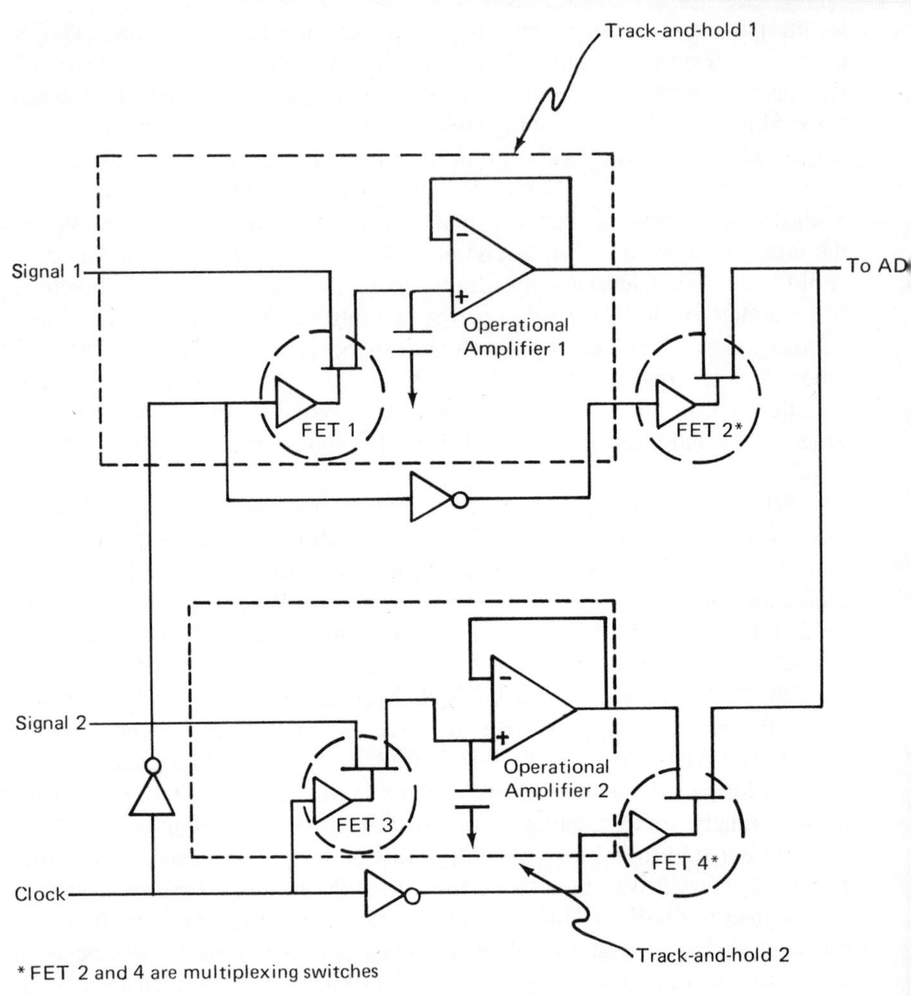

Fig. 11-2
Two-channel multiplexer with track-and-hold amplifiers.

by sampling two simultaneous transient analog signals by switching them alternately into a single digital input channel. Figure 11-3 shows the flow chart for a typical multiplexing data-acquisition application using the computer for storage and data processing. Figure 11-4 shows a REAL-TIME BASIC program to implement the flow chart.

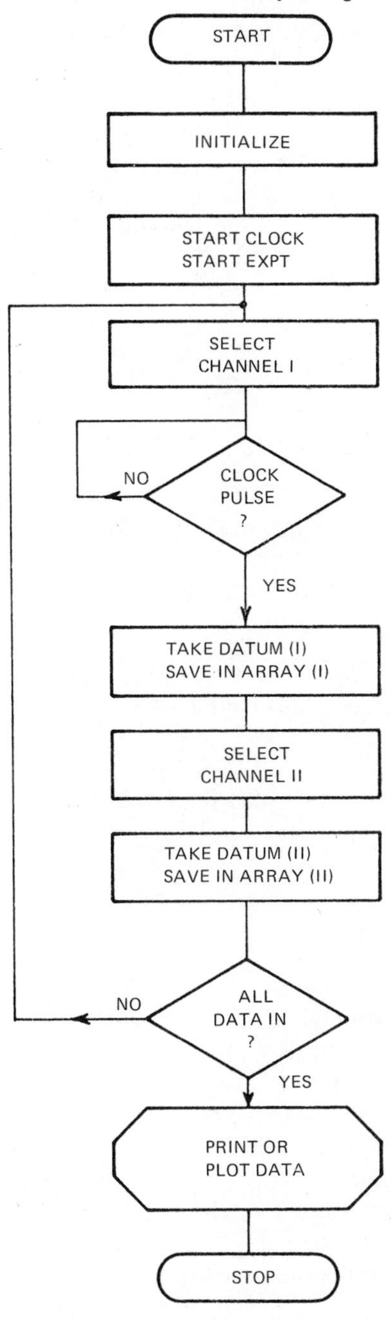

Fig. 11-3
Flow chart for two-channel multiplexed data acquisition.

	PROGRAM	COMMENTS
2	DIM M(1ØØ)	
3	DIM Q(1ØØ)	
5	DIM D(2)	
6	LET C = 17	MPXR Control I/O channel number = 17
1Ø	LET F = 1.0	Clock freq. = 1.0 Hz
15	CALL INIT (F)	Initialize clock
2Ø	CALL START	Start clock
25	FOR N = 1 TO 1ØØ	
3Ø	FOR I = 1 TO 2	
4Ø	CALL DIGOUT (I, C)	Select MPXR channel
5Ø	CALL ADC(X)	Get datum
6Ø	LET D(I) = X	
7Ø	NEXT I	
8Ø	LET M(N) = D(1)	Store two data in two arrays
9Ø	LET Q(N) = D(2)	
1ØØ	NEXT N	
2ØØ	CALL STOP	Stop clock
	⋮	
3ØØ	END	(Print, plot, etc.)

Fig. 11-4

Sample REAL-TIME BASIC program for two-channel multiplexing according to flow chart of Fig. 11-3.

EXPERIMENTAL

Equipment Needed

1 EL Digi-designer
1 ±15-V power supply
1 Hex inverter package (SN 7404)
4 Field-effect-transistor switches (CAG-30)
2 BCD counters (SN 7490)
2 Operational amplifiers (SN 72741)
2 0.01-μF capacitors
1 Pulse generator (variable rate)
1 Dual-trace oscilloscope
1 Dual-function generator (see Appendix E), 10-Hz triangular and square waveforms
1 Laboratory computer system with programmable clock and digital and analog I/O interfaces

Procedure

A. Computer Control of Data Acquisition by Multiplexing

1. Write a program to acquire data, to direct the data from each input signal to the correct storage block, to produce the necessary control signals, and finally to retrieve and plot the data for each on the storage oscilloscope or Teletype.

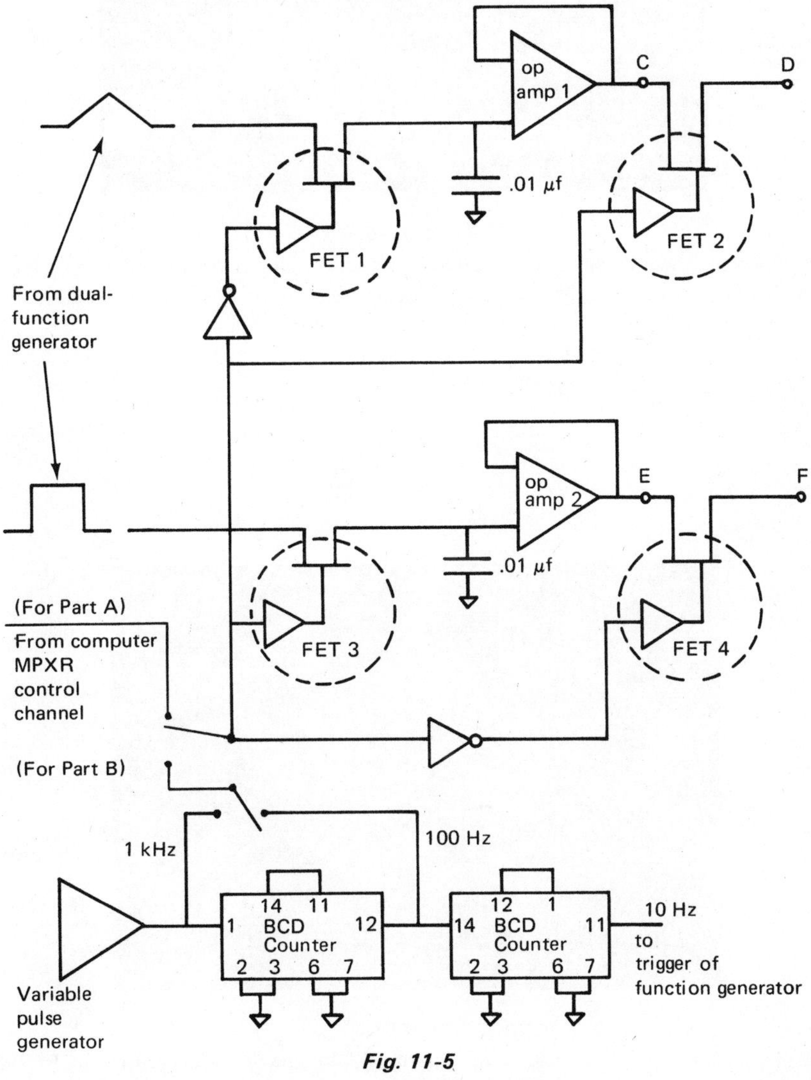

Fig. 11-5
Simple two-channel multiplexer.

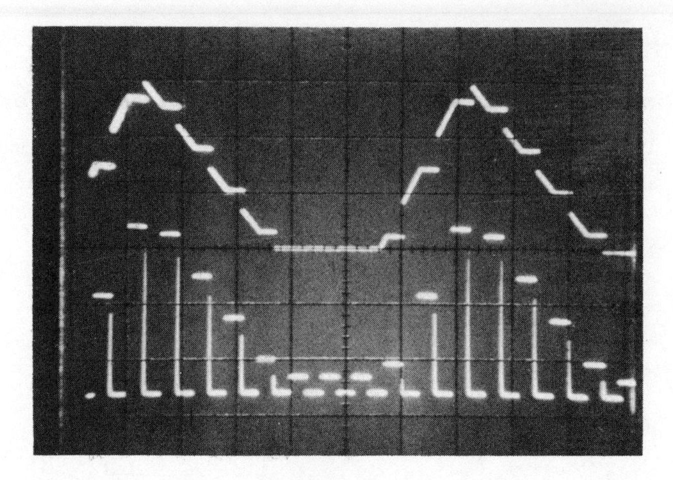

Fig. 11-6

Oscilloscope trace. Upper trace—output of op amp 1; lower trace—output of FET-2 (see Fig. 11-5). Time base, 10 msec/div; vertical scale, 1 V/div. Experiment as in Sec. B1.

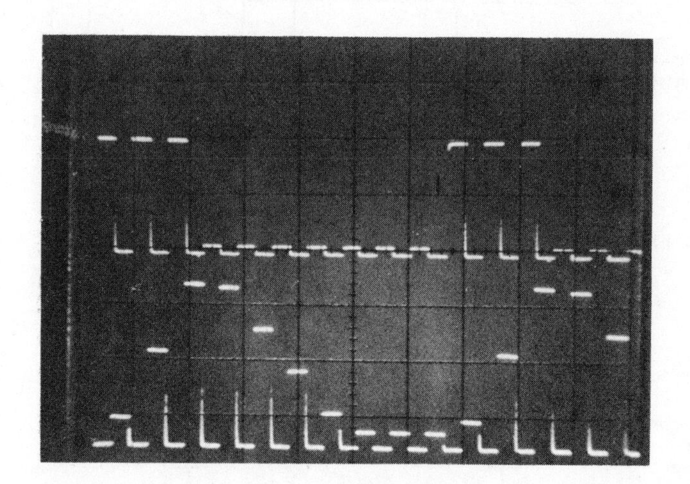

Fig. 11-7

Oscilloscope trace. Upper trace—output of FET-4; lower trace—output of FET-2 (see Fig. 11-5). Time base, 10 msec/div; vertical scale, 1 V/div. Experiment as in Sec. B3.

2. Wire the computer system, dual-function generator, and multiplexer in Fig. 11-5. (Alternately, you may use the multiplexer on the computer, if available, instead of wiring that in Fig. 11-5.) Run the program.

B. *Two-Channel Multiplexer (Optional)*

1. Consult the data sheets or pin diagrams for the CAG-30, the SN 72741 operational amplifiers, the SN 7404 inverter, and the counters. Power them correctly and wire the circuit shown in Fig. 11-5. Use the 10-Hz output of the counter to trigger the function generator. Connect channel 1 of the oscilloscope to point *C*, and channel 2 to point *D*. Set the pulse-generator frequency to 1 kHz, the oscilloscope to 10 msec/div. horizontal and 1 V/div. vertical. Does FET switch 2 turn "on" when the T/H amplifier is in the "hold" mode, i.e., FET-1 is off? Refer to Fig. 11-6 for a photograph of a typical oscilloscope trace obtained for this part of the experiment.
2. Repeat the procedure above, following the square-wave signal this time.
3. Connect channel 1 to the output of the triangular wave line at point *D*, and channel 2 to the square wave line at point *F*. Does the multiplexer switch to each line alternately? Refer to Fig. 11-7 for a photograph of a typical oscilloscope trace for this part.

Experiment 12

Determination of Data-Acquisition Rate — The Nyquist Frequency*

In many applications of laboratory computers, it is necessary to sample a periodically varying analog signal and to store its digitized representation in the computer. In general, such data is meaningful only if certain conditions are met during the acquisition phase. The purpose of this experiment is to illustrate the principles involved in meeting these conditions. A more detailed discussion, including a detailed error analysis, is available in ref. D-4.

Determination of the minimum usable data-conversion rate is an experiment most often ignored, and therefore errors introduced by lack of data go unnoticed. Some possible errors are:

1. Distortion of data waveforms
2. Excess noise levels
3. Actual modification of frequencies

It is easiest to discuss the above kinds of errors in terms of frequencies, and since the waveforms are rarely known in advance, a signal is best described in terms of its sine and cosine wave components. This is especially true in cases where measurement of a waveform is the object of the experiment. For this purpose, it is sufficient to examine the power content at each frequency and not be especially concerned with phase information. The mathematical process of calculating the required power spectra is the Fourier transform. Samples are given in Fig. 12-1. The minimum data-acquisition rate which can be used is the one which captures all of the frequency information in the unknown signal.

For a given choice of sampling rate, there is a maximum periodic signal frequency which can be correctly sampled; signals with frequencies less than

*Figures and parts of the discussion reprinted with permission from R. Swanson, D. J. Thoennes, R. C. Williams, and C. L. Wilkins, *J. Chem. Educ.*, **52** (1975).

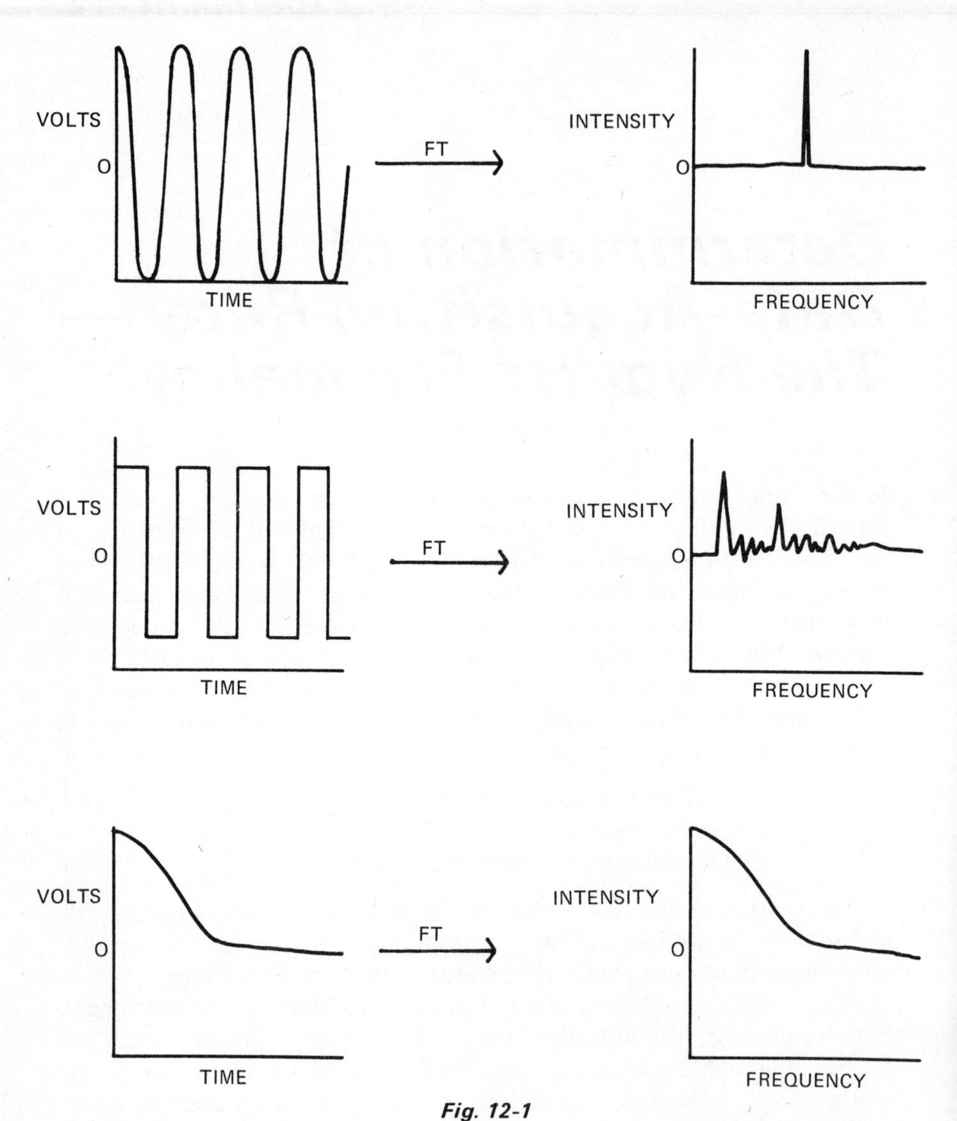

Fig. 12-1
Original and Fourier-transform-calculated (FT) power contents of various signals.

this value will generally be measured correctly, while those with higher frequencies will *appear*, after sampling, to have frequencies below the critical frequency. In the latter case, the "measured" amplitude may be any value between zero and the maximum signal amplitude. Conversely, if the periodic signal frequency is fixed and the sampling rate is varied, there is a minimum rate of sampling which must be used for sampled points to correctly represent the signal. A sampling of the signal below this threshold rate will result in distortion of the waveform to produce an *apparent* signal of lower frequency (and possibly lower amplitude) than the actual signal. The distortion observed in both cases is called "aliasing." *Aliasing occurs whenever the sampling rate is less than twice the signal frequency.* The frequency at which distortion or aliasing begins is called the Nyquist frequency,

$$N_{\text{signal}} = \text{sampling rate}/2 \tag{12-1}$$

$$N_{\text{sample}} = \text{signal frequency} \cdot 2 \tag{12-2}$$

and its definition depends on whether it is the sampling frequency or the signal frequency which is fixed. Thus, the *Nyquist signal frequency* is the *maximum* signal which can be correctly digitized at a given sampling rate, while the *Nyquist sample frequency* is the *minimum* sampling rate necessary if a given signal is to be correctly reconstructed from the samples.

ALIASING

An understanding of the aliasing phenomenon is vital to the use of the new computer-assisted experimental methods now available. For example, in Fourier-transform nuclear-magnetic-resonance experiments, an incorrect choice of signal sampling rates can lead to spurious peak assignments. This is shown in Fig. 12-2 for the low-resolution proton NMR spectra of mesitylene (1,3,5-trimethylbenzene). Both spectra were taken on a Varian XL-100 spectrometer operating at 100 MHz in the Fourier-transform mode. In both cases, the effective data-acquisition rate was 2000 Hz, yielding a Nyquist signal frequency of 1000 Hz. In the upper trace, the transmitter frequency was chosen so that the TMS resonance appeared at 176 Hz, and the mesitylene aromatic band appeared at 852 Hz (676 Hz higher frequency than TMS). In the lower trace, the transmitter frequency was chosen so that TMS peak was detected at 375.5 Hz. Since the mesitylene aromatic resonance is 676 Hz higher, it should occur at 1051.5 Hz, which is above the Nyquist signal frequency. The measured result is "folded" back below the Nyquist signal frequency and appears to indicate a band at 948.5 Hz (574 Hz above TMS), over 1 part per million in error.

Although a detailed discussion of Fourier-transform techniques is beyond the scope of this article, the aliasing phenomenon itself is relatively simple. Consider first the case of a fixed signal frequency sampled at several digitization rates. This is illustrated in Fig. 12-3. The sine wave in Fig. 12-3a represents the actual analog signal, which has a frequency of $\frac{1}{2}$ Hz (and, therefore,

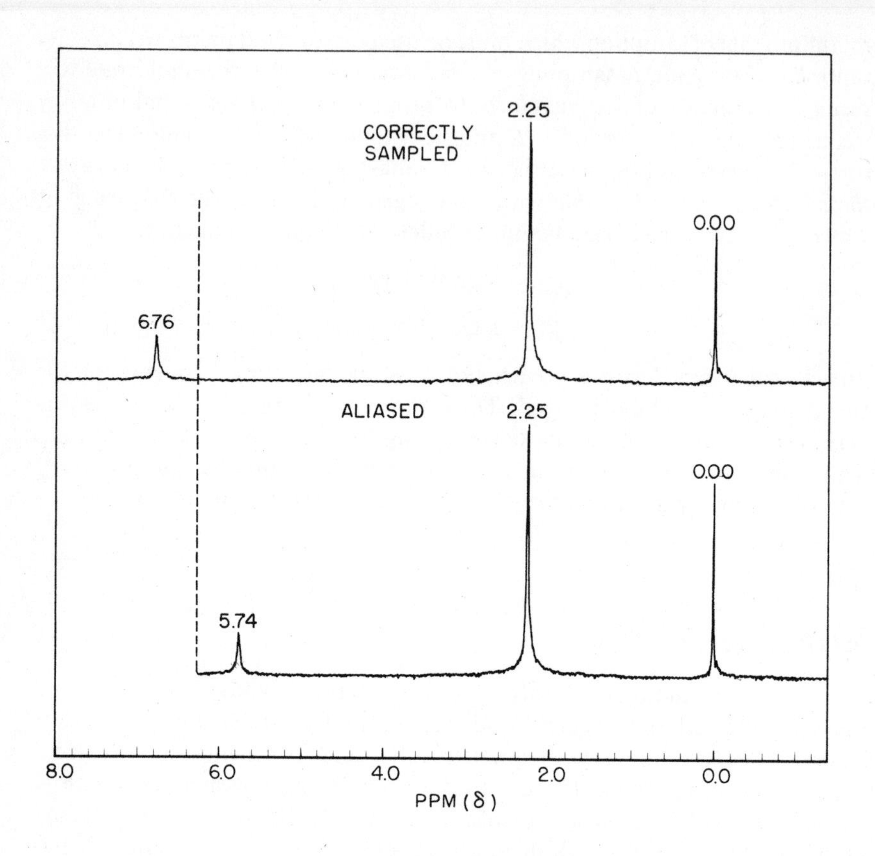

Fig. 12-2

Low-resolution proton NMR spectra of mesitylene measured in FT mode on a 100-MHz instrument (mesitylene 10% in $CDCl_3$, 1% TMS). The dotted line represents the maximum frequency which can be correctly sampled, for the lower trace. Upper trace: correctly sampled spectrum. Lower trace: incorrectly sampled spectrum which "folded" back, yielding an aliased result.

a Nyquist sample frequency of 1 Hz). The tic marks on Fig. 12-3a show when the signal is digitized for the various sampling rates used in Fig. 12-3b to 12-3e. The effects of sampling at less than the Nyquist sample frequency are shown in Figs. 12-3b, 12-3c, and 12-3d. When the sampling rate is $\frac{1}{3}$ Hz (Fig. 12-3b), the signal has the correct amplitude, but appears to have a frequency of $\frac{1}{3}$ Hz. (The exact agreement between the sampling rate and the apparent signal is coincidental, but the resultant aliased frequency will always be less than, or equal to, the sampling frequency.) Setting the sampling

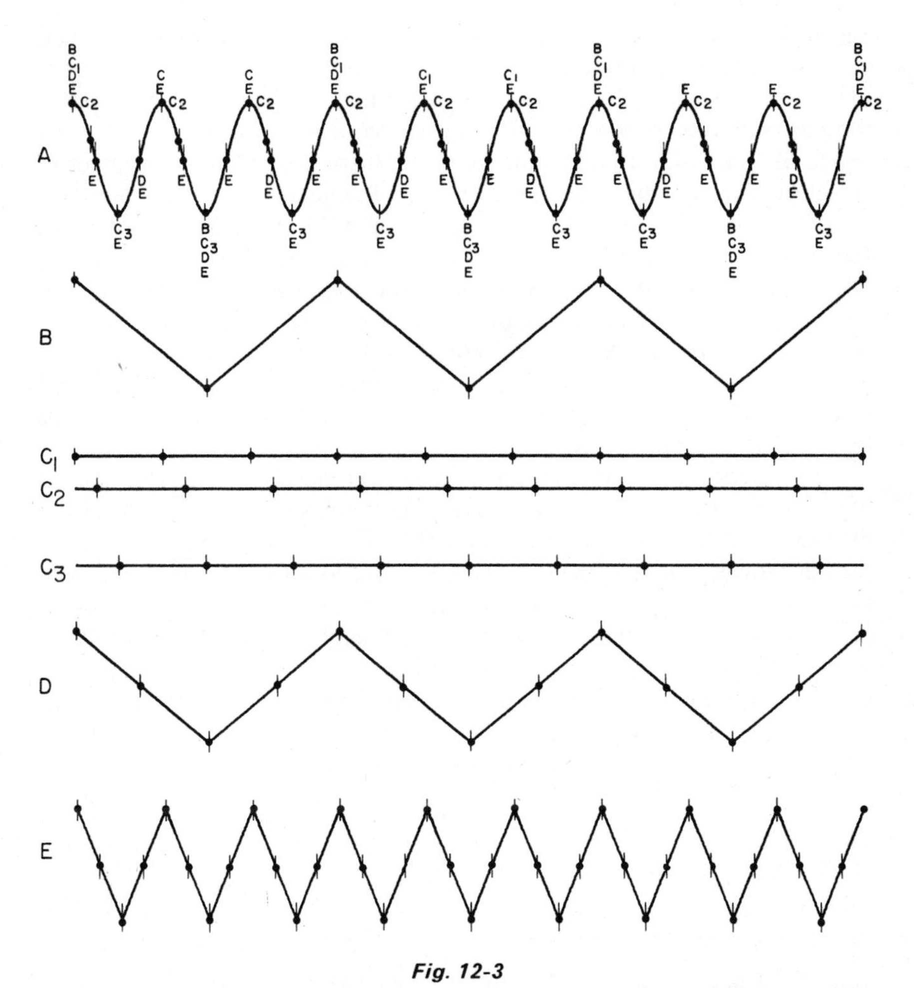

Fig. 12-3

Sampling of a $\frac{1}{2}$-Hz sine wave at various frequencies. (a) Actual signal; (b) sampling at $\frac{1}{3}$ Hz; (c) sampling at $\frac{1}{2}$ Hz; (d) sampling at $\frac{2}{3}$ Hz; (e) sampling at 2.0 Hz. Letters on curve a show sample points for respective curves whose letters are given.

rate equal to the signal frequency (Fig. 12-3c) yields even worse results. Note that an apparent frequency of zero is obtained and that the position of the resultant line depends on where in the signal-frequency cycle the sampling is begun. Sampling at $\frac{2}{3}$ Hz (Fig. 12-3d) does not yield correct results either, because the frequency is still less than the Nyquist frequency (notice that the *apparent* signal frequency is again $\frac{1}{3}$ Hz). Finally, sampling at 2 Hz (Fig. 12-3e), a rate greater than the Nyquist sample frequency, yields both the correct amplitude and the correct frequency. It is *never* possible to obtain the correct signal data unless the sampling rate is at least equal to the Nyquist frequency.

The agreement of the aliased frequencies in Figs. 12-3b and 12-3d is not uncommon; for a given signal frequency there are an infinite number of sample rates (below the Nyquist sample frequency) which give the same result. Similarly, for a given sampling rate, there are an infinite number of signal frequencies (above the Nyquist signal frequency) which give the same aliased result. This is shown graphically in Fig. 12-4. Here, the x axis denotes the signal frequency, where N_S is the sampling frequency, and the y axis denotes the digitized result. So long as the signal frequency is less than $N_S/2$, the correct frequency is obtained. For signal frequencies between $N_S/2$, and N_S, the aliased result *decreases* with increasing signal frequency and may exhibit different phase characteristics. Thereafter, the measured result will increase between nN_S and $(nN_S + N_S/2)$ and decrease between $(nN_S + N_S/2)$ and $(n + 1)N_S$.

In practice, a variable signal frequency is usually sampled at a fixed digitization rate, as in Fig. 12-4. An example of aliasing resulting from such a fixed sampling and variable signal rate is the "wagon-wheel" effect commonly seen in movies, where a spoked wheel appears to move forward, stop, and move backwards as the wagon accelerates; occasionally, this process can be observed several times before the spokes finally appear to reach a state of equilibrium. In this instance, the sampling rate is the number of picture frames exposed in one second, and the signal rate is defined by the speed and

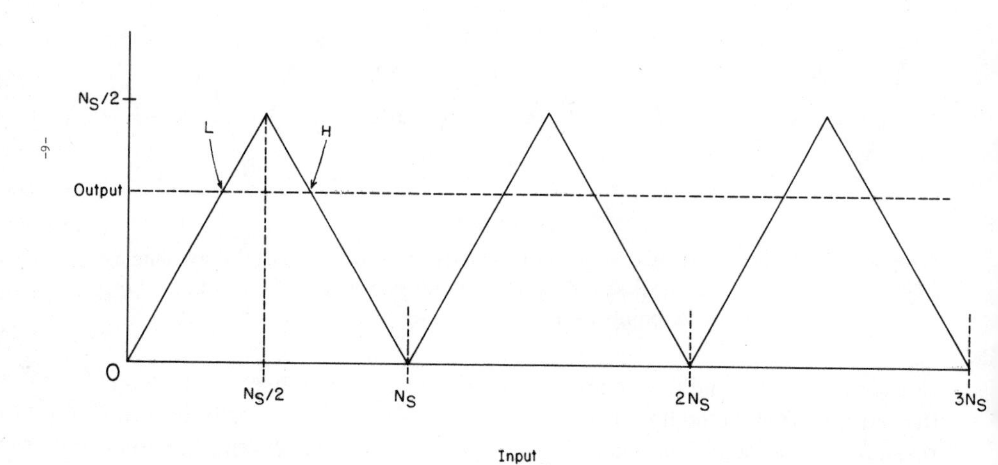

Fig. 12-4

A plot of actual (input) frequency vs apparent (output) frequency at a fixed sampling rate N_S. L and H represent places where the same *apparent* frequency results from signals of different frequencies.

number of spokes in the wheel. As long as it is possible to take at least two pictures before a new spoke attains the orientation of a previous one, the correct wheel speed will be recorded on film. When it is no longer possible to do this (i.e., the wheel-spoke frequency is greater than the Nyquist signal frequency), the signal is aliased, and the wheel appears to go backward (the phase difference mentioned earlier). When the spoke frequency and the sampling rate are equal, the spokes in the wheel appear motionless. If the wheel continues to accelerate, the entire cycle will be repeated.

There are two additional points which are illustrated by this example. The first is that if aliasing occurs, it is not possible to tell, *from the final picture*, the frequency of the wheel rotation; i.e., once the data is aliased, it is usually lost for good. Second, while two frames per spoke cycle are sufficient to record the wheel speed correctly, they may not be sufficient to preserve the sensation of smooth, continuous motion. This is also shown in Fig. 12-3. Thus, when digitizing periodic functions, sampling at or just above the Nyquist frequency preserves only the frequency information of the signal, not the shape. If it is necessary to record the signal shape, the sampling frequency must be several times the signal frequency. Although the frequency information itself is often sufficient, it is good practice to sample at a rate well above the minimum required. This both avoids possible aliasing when the signal frequency is somewhat higher than expected and leaves the option of reconstructing the signal shape.

The present experiment is designed to establish the relationship between the analog-to-digital conversion rate and the frequency of a sine-wave voltage input. This experiment permits the student to experimentally determine the maximum allowable signal frequency (i.e., the frequency N_{signal} above which periodic signals will be "folded" or "aliased" to lower frequencies). The method used is to sample a suitable standard sine-wave signal source with an analog-to-digital converter operating at a single fixed sampling rate (the Nyquist sampling frequency). These data are stored in the computer. Then the data thus acquired are converted to a convenient audiofrequency by "playing back" at an appropriate (usually higher) rate via an audio amplifier and speaker combination. For a given setting of the digitization rate, if the input sine-wave frequency is varied in steps of 20 Hz from 60 to 400 Hz, a series of ascending tones, followed by a blank series (where the output tone is above 10,000 Hz) and then a series of descending tones, will be heard. The descending tones are a result of the higher frequencies being *aliased* to lower frequencies, once N_{signal} is passed. Thus, direct audible observation of aliasing is possible. N_{signal} can be determined aurally by selecting *two* input sine-wave frequencies, one below and one above (points L and H in Fig. 12-4) the estimated Nyquist signal frequency, and adjusting the higher frequency until the tone matches that produced by the lower. (It is

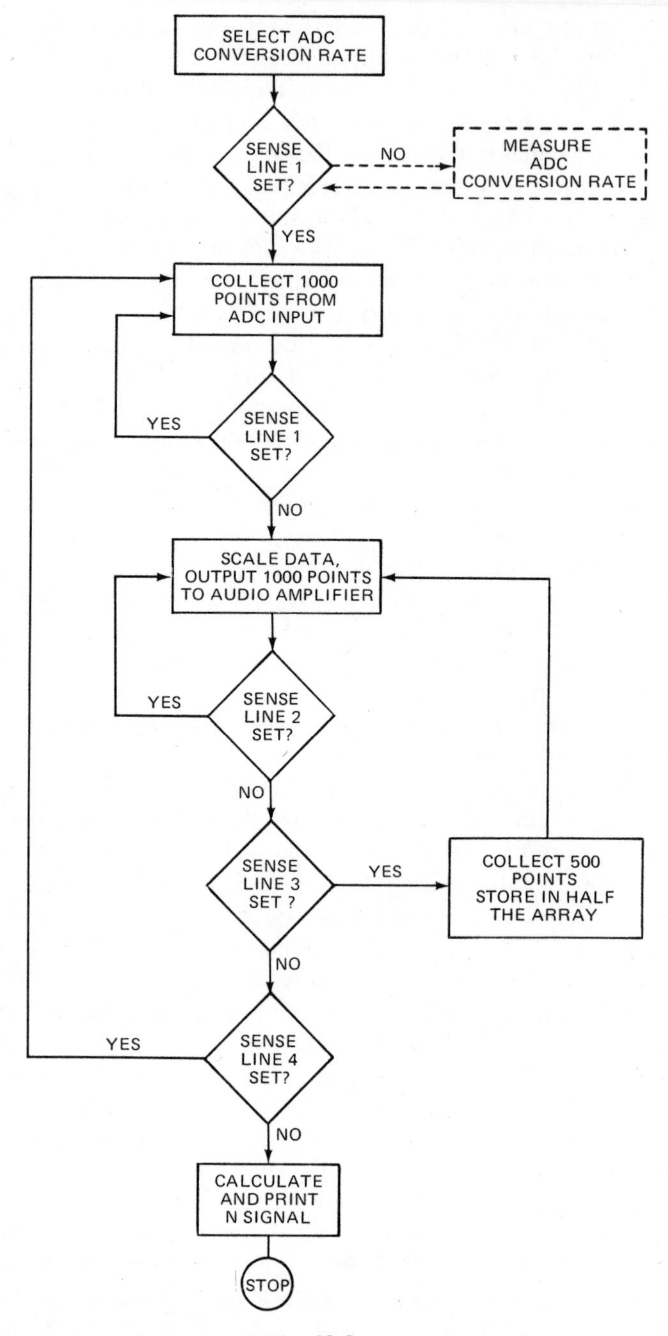

Fig. 12-5

 Flow chart of a sample program for the Nyquist frequency experiment.

equally possible to fix the higher frequency and vary the lower.) When the two frequencies are played back sequentially, most human ears can distinguish mismatches as small as 0.5 Hz. N_{signal} is the average of the two frequencies. The experimentally determined value of N_{signal} should, in every case, be very close to half the known sample digitization rate (N_{sample}). The experiment should be repeated for several fixed digitization rates to verify that this is so.

A flow chart of a sample program is shown in Fig. 12-5. An optional portion is enclosed in broken lines. This program has been designed to provide a number of controls so that three related experiments can be carried out to demonstrate the aliasing phenomenon when the ADC conversion rate is less than the Nyquist sampling frequency. The program flow chart provides for the input into an array of one signal; then, depending upon the status of computer sense lines or panel switches, the program outlined will either output (play back) the array, enter a new signal into the entire array, or enter a new signal into the last half of the array. If the entire array is played back after filling the last half of it with a new signal, then the first half of the playback corresponds to the old signal, and the second half corresponds to the new signal. Any difference in the output frequencies appears as a change in the tone of the audio playback, occurring halfway through the playback. Although the flow chart specifies a 1000-point array, it is useful to increase the size of this array, if more core memory is available, so that the audio playback time is increased.

In an experiment with real data, the frequency distribution of noise must not be neglected when setting up an experiment. In fact, the most important modification of the Nyquist criterion is proper selection of filters between the experiment and the analog–digital converter (ADC). Consider the data and power spectrum given in Fig. 12-6. The power spectrum indicates that little real data exist above frequency F_1, and therefore a digitization rate of $2F_1$, or F_2, is sufficient. Notice, however, that noise exists at all frequencies, and due to the "folding back" discussed above, noise between frequencies F_1

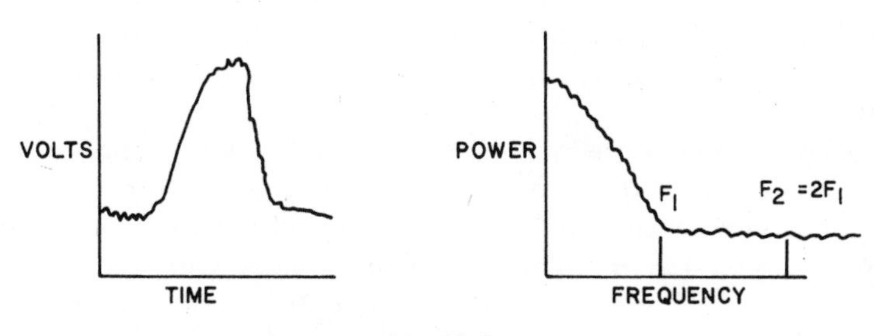

Fig. 12-6

Power spectrum of real data.

and F_2 will be added to the noise already existing between 0 and F_1. Once digitized, it is impossible to distinguish the added noise from the signal-frequency noise, and therefore the overall signal-to-noise ratio will have been degraded. If, however, an appropriate analog filter is used to remove all frequencies above F_1 *before* digitization, the information content of the digitized signal will be enhanced.

EXPERIMENTAL

Equipment Needed

1 Variable sine-wave source (6–600 Hz)
1 Audio amplifier and speaker (100–10,000 Hz)
1 Dual-trace oscilloscope
2 Resistance substitution boxes
2 Capacitance substitution boxes
1 Laboratory computer system with programmable clock and digital and analog I/O interfaces

Procedure

Students should write a program from the flow chart in Fig. 12-5. If it is possible to do so with the laboratory computer system used, the program should include a provision for the direct measurement of the ADC conversion rate using an oscilloscope. This is particularly true for ADC's whose conversion times are slow and vary as a function of the input voltage. For the relatively slow integrating ADC, faster conversions are achieved for small input voltages. In this case, it may be necessary to adjust the sine-wave signal voltage so that the time for a single analog-to-digital conversion is less than the desired acquisition rate. This procedure may not be necessary if a faster successive-approximation type ADC is used; the laboratory instructor should specify the type used, and whether this step is necessary. The program should use student-set panel switches or sense lines to control data acquisition and playback, to indicate when a frequency match has been found, and to govern other branching decisions which students desire to control externally.

After the program is written, students should set up the hardware as shown in the block diagram of Fig. 12-7. The output-filter-circuit block diagram is shown below in Fig. 12-8. A value of about 4700 Ω for R_1 and between 0.1 and 0.22 μF for C_1 should be satisfactory for output signals in the range 1–2 V. This filter circuit is necessary to smooth out the "steps" in the tone coming through the speaker from the digital-to-analog converter

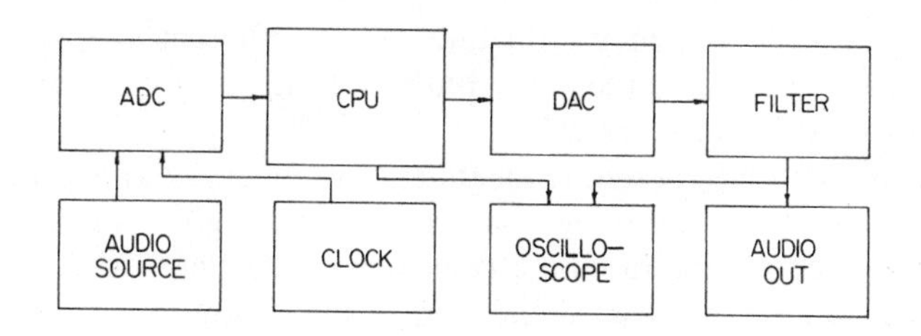

Fig. 12-7
Block diagram for Nyquist-frequency-experiment hardware.

(DAC). The program should be used to acquire and play back data for one frequency and, if necessary, measure the ADC conversion rate. Students should use an oscilloscope, if available, to observe the unfiltered and filtered output of the DAC for several different combinations of R_1 and C_1. This process should be repeated for several different signal frequencies above and below the Nyquist signal frequency.

Finally, after storing the signal frequency *below* N_S in the *entire* data array, a signal frequency *above* N_S is stored in the *second* half of the array, and the entire array is subsequently played back. (As was discussed earlier, the array now consists of two different frequencies—the lower frequency in the first half, and the higher frequency in the latter half of the array.) If the tones match, students should enter the numerical values of the frequencies (as read from the sine-wave signal generator) so that N_{signal} may be calculated and printed.

If no match of tones is obtained on the first trial, a new high-frequency sine wave should be generated and input into the second half of the array, and the entire array should again be played back. (The first half of the array still contains the low-frequency sine wave.) This process is continued until a

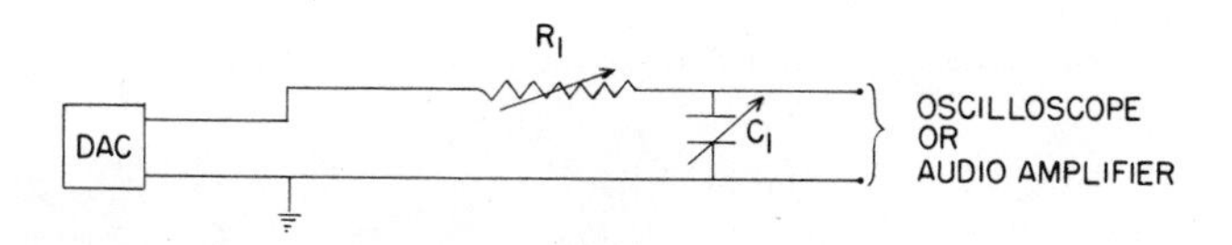

Fig. 12-8
Filter circuit for audio playback.

match is ultimately obtained. The entire procedure should then be repeated for several different frequency pairs and several different acquisition rates. (Best results are achieved if the audio output of the lower frequency is in the middle of the human audio range.)

As an alternate approach, students may want to use this procedure to calculate the minimum sampling rate necessary for a given signal frequency by varying the acquisition rate for a fixed signal frequency. Also, if time permits, an analog filter, similar to the output filter, should be designed and placed between the signal source and the ADC. The experiment described above can then be repeated to demonstrate that the best signal-to-noise ratio is obtained by removal of aliased noise.

Computer-Aided Demonstration of the Photoelectric Effect

In 1887, Hertz observed that when certain materials were irradiated with ultraviolet radiant energy, they gave off electrons. This phenomenon is called the photoelectric effect. Figure 13-1 shows a simple diagram of a photoelectric-effect setup. A small voltage is impressed (usually 0–50 V) across the cold electrodes in order to accelerate electrons away from the photoconducting surface. The galvanometer G shows the magnitude of the current flowing in the circuit at any time.

Several interesting facts were found. The first of these is that no electrons were ejected from the irradiated electrode unless a certain minimum energy (or wavelength) was used in the irradiation. The second is that the number of electrons ejected (current) depended upon the intensity of radiation; indeed, a linear relationship between radiation intensity and photocurrent was found. Third, the speed (kinetic energy) of the emitted electrons was dependent on the wavelength of radiation energy used and independent of the intensity of irradiation.

In 1905, Einstein explained these facts by applying his quantum law. The energy of a photon is $h\nu$, and for an electron to be ejected from a substance, this energy must equal the work function W (the minimum energy required to cause electron ejection where the electrons are ejected with no kinetic energy, KE). The quantity h is Planck's constant, and ν is the wavelength of the incident radiation. Equation (13-1) summarizes this relation.

$$h\nu = W + KE \tag{13-1}$$

It is obvious that, regardless of intensity, no electron ejection can occur until the photon contains at least W energy. If the photon energy is larger than W, the remainder will be converted into kinetic energy. This result was verified by Robert A. Millikan in 1911.

If Eq. (13-1) is rearranged, it is seen that the kinetic energy KE is equal to

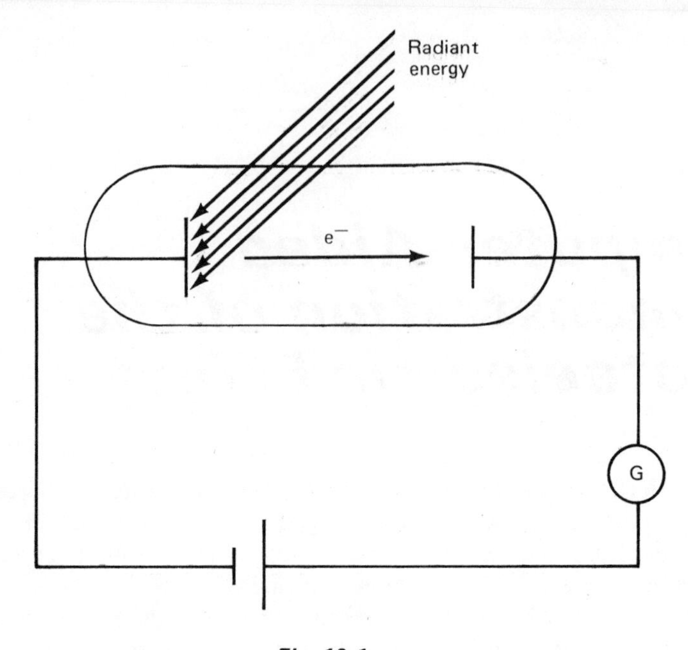

Fig. 13-1
Schematic diagram of the photoelectric effect.

the difference of the photon energy $h\nu$ and the work function W. If a potential V_0 just sufficient to prevent current flow is applied across the electrodes

$$KE = h\nu - W \tag{13-2}$$

$$eV_0 = h\nu - W \tag{13-3}$$

$$\nu = \frac{eV_0}{h} + \frac{W}{h} \tag{13-4}$$

that energy must equal KE, and thus the product of the charge of an electron ($e = 1.6021 \times 10^{-19}$ C) and the applied potential V_0 may be substituted, yielding Eq. (13-3). Rearranging, Eq. (13-4) results.

If several different optical filters are used with the source, and the "stopping potential" V_0 is plotted on the abscissa versus ν on the ordinate, then a straight line with a slope of e/h and an intercept of W/h should result. It should be noted that since the currents being monitored are very small, special care must be given to the amplification and scaling factors used in the computer and any external hardware. The range of amplification necessary should be large enough so that the maximum current value will produce nearly a full-scale reading from the ADC but within the limits of the recorder.

Since the currents are of the order of micro- to picoamperes, it is necessary to amplify it by a factor of 10^8 or 10^9. The exact amplification required should be determined prior to connecting the experiment with the computer.

This experiment is designed to introduce you to the problem of making measurements using a very sensitive circuit (currents of the order of 10^{-10} A) and to increase general experience in the use of computers on laboratory problems.

EXPERIMENTAL

Equipment Needed

1 Vacuum photodiode in lighttight housing. (This should be constructed with a provision for illumination of the photocathode.) It can be made from parts such as an RCA 929 tube and a lighttight box painted black
1 Light source (such as mercury arc)
1 Set of optical filters, preferably narrow-band filters
1 Electrometer (such as a Keithley 911)
1 Laboratory computer system with programmable clock and digital and analog I/O interfaces

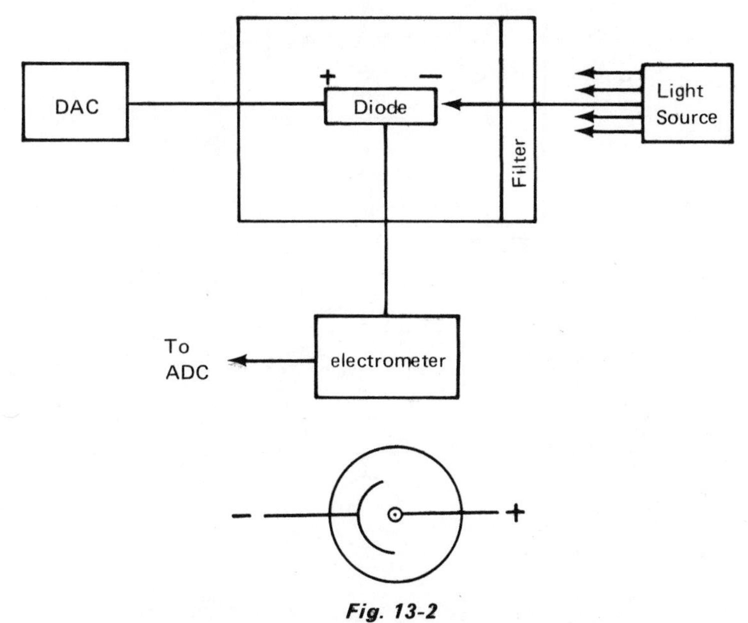

Fig. 13-2
Block diagram for photoelectric-effect experiment.

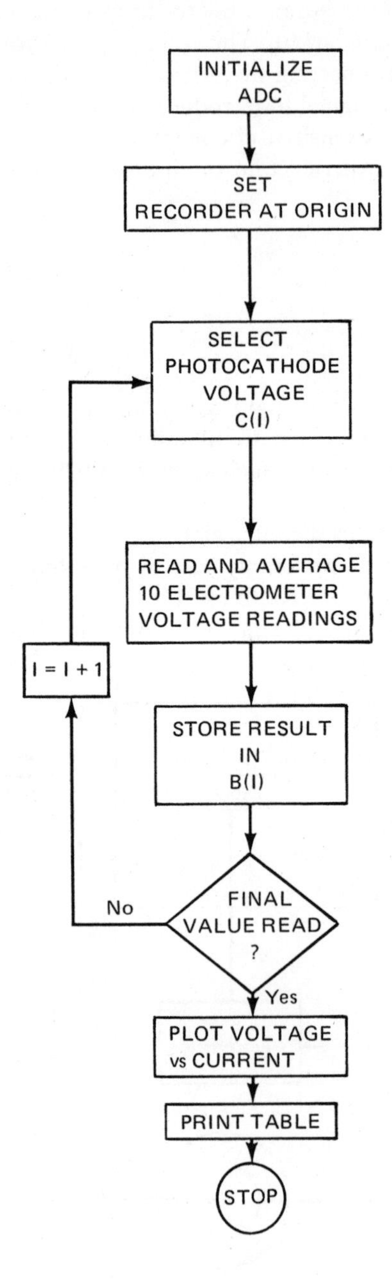

Fig. 13-3
Block diagram of photoelectric-effect measurement program.

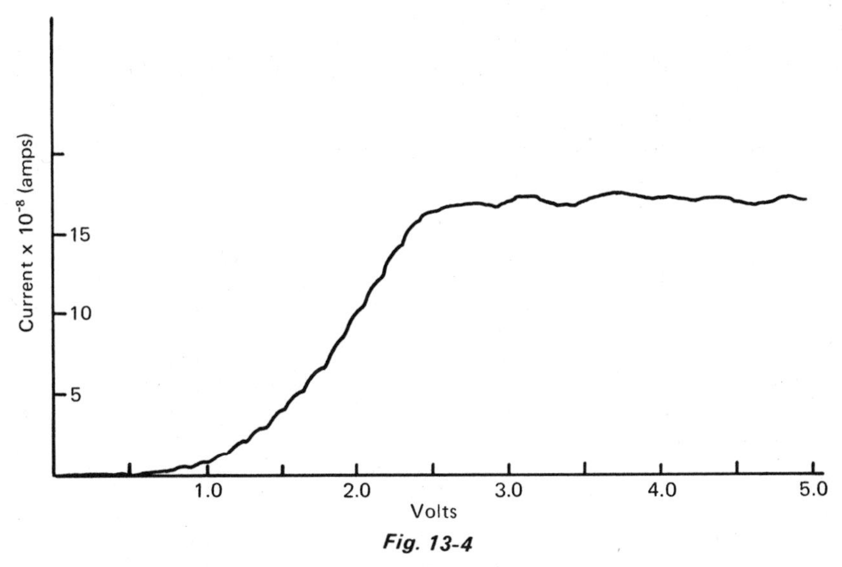

Fig. 13-4

Current vs. voltage for a typical photoelectric effect.

Procedure

1. Wire the circuit as shown in Fig. 13-2. The vacuum photodiodes can be any of several commercially available tubes. Generally, they will have four or five pins of which only two are connected. Consult the appropriate diagram to determine which pins should be connected. Tube manuals will also show the general spectral-response curves for the photodiodes.
2. Write a program which will produce and control the reverse voltage on the photocathode (range 0–5 V) and measure the output voltage of the electrometer which corresponds to current from the photodiode. Figure 13-3 shows a block diagram of a program which performs these functions. Since each of the readings may contain random noise, 10 readings are averaged at each applied voltage. The program *may* be designed to save the stopping potentials and ultimately calculate the e/h value by applying a least-squares calculation. Figure 13-4 shows a typical plot of data obtained, and Table 13-1 shows the actual data corresponding to this plot.
3. Plot the electrometer output voltage as a function of current.
4. Run the program and plot the data using several different filters corresponding to different wavelengths striking the photocathode.

Table 13-1

Sample Output Data from A Photoelectric-Effect Experiment

Filter 22 Volts	Reading Amps × 10^8	Filter 22 Volts	Reading Amps × 10^8
.125	−0.4	2.265	16.34
.25	−0.4	2.75	16.686
.375	−0.4	2.875	16.728
.5	−0.376	3	16.49
.625	−0.302	3.125	17.076
.75	−0.18	3.25	17.15
.875	−0.004	3.375	16.684
1	0.26	3.5	16.632
1.125	0.594	3.625	17.144
1.25	1.182	3.75	17.426
1.375	1.922	3.875	17.302
1.5	2.844	4	17.148
1.625	3.98	4.125	17.156
1.75	5.376	4.25	16.938
1.875	6.934	4.375	17.184
2	8.86201	4.5	17.158
2.125	10.628	4.625	16.858
2.25	12.432	4.75	16.898
2.375	14.468	4.875	17.342
2.5	15.908	5	17.12

5. Utilizing the information obtained in step 4, calculate the value of e/h.
6. *Optional*: Determine the work function W of the photocathode. (Work function and ionization potential are synonymous.)

Experiment 14

Computer-Assisted Potentiometric Titrations

The Nernst equation is the basis for carrying out potentiometric measurements. For a generalized redox reaction

$$aA + bB \rightleftarrows cC + dD \tag{14-1}$$

the potential of the cell is given by

$$E = E^\circ_{\text{cell}} - \frac{RT}{nF} \ln \frac{(a_C)^c (a_D)^d}{(a_A)^a (a_B)^b} \tag{14-2}$$

Separating it into two half-reactions, we have

$$E_1 = E^\circ_1 - \frac{RT}{n_1 F} \ln \frac{(a_C)^c}{(a_A)^a} \tag{14-3}$$

and

$$E_2 = E^\circ_2 - \frac{RT}{n_2 F} \ln \frac{(a_D)^d}{(a_B)^b} \tag{14-4}$$

For any system which is not at equilibrium, there will be a finite potential difference between the two half-reactions. This potential difference can be measured with a *potentiometer*. A potentiometer is a simple device which employs a standard source, such as a Weston cell with a known potential, as one element in a bridge circuit. The unknown potential is also placed in the circuit, and the two are balanced against each other, using a voltage divider, until no current flows, as shown by the galvanometer *G*. Then, from the value of the known potential and divider factors, the ratio of standard to unknown voltage and, hence, unknown potential can be calculated. Figure 14-1 shows such a simple circuit. In actual practice, the potentiometer used is more complex than the one whose diagram is shown in Fig. 14-1. For more

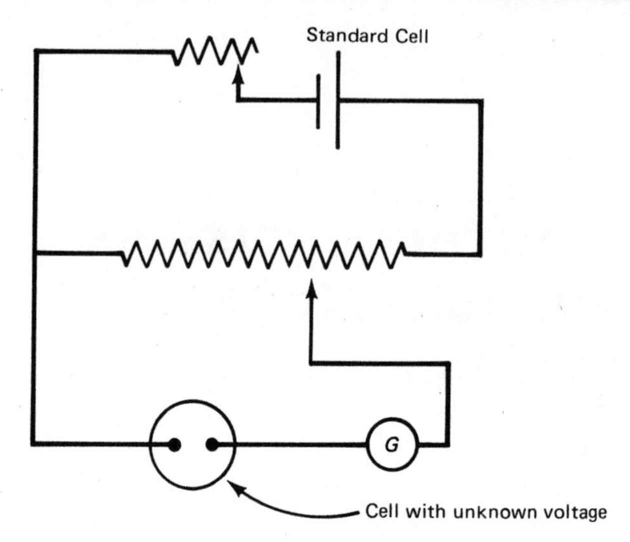

Fig. 14-1
A simple potentiometer circuit.

detailed discussions, you should refer to appropriate texts containing chapters on potentiometric measurements.

In potentiometric titrimetry, the potential of an electrode pair is followed as a function of the volume of titrant added. One of the electrodes (the reference electrode) has a constant potential. Therefore, the potential change observed, as titrant is added, is the result of the change in the other electrode (the indicator electrode). The reference electrode may be a saturated calomel electrode (S.C.E.). The indicator electrode could be inert (like platinum), noninert (silver, mercury), or a specific ion electrode (H^+, Ca^{++}, F^-, etc.). Its potential must be dependent upon the activity of one or more of the species involved in the titration reaction.

As titrant is added, the potential of the reference-indicator pair will change only slowly as the concentration of the titrate slowly changes. Table 14-1 shows such changes. As the equivalence point is approached, the concentration of the titrate changes very rapidly for a given volume of titrant, and the potential of the reference-indicator pair also changes very rapidly. As the equivalence point is passed, the concentration or activity once more changes slowly. This results in the familiar S-shaped titration curve.

In acid–base titrations, the pH-sensitive glass electrode is used as the indicator electrode. The glass electrode employs an internal silver wire

Table 14-1

Change in Titrate Concentration/0.10 ml Titrant Added*

Total volume titrant added (ml)	$[H^+]$	Δ[Conc.]	% change
0.00	0.1000		
0.10	0.0996	0.0004	0.40%
20.00	0.0429		
20.10	0.0427	0.0002	0.47%
30.00	0.0250		
30.10	0.0248	0.0002	0.80%
40.00	0.0111		
40.10	0.0110	0.0001	0.90%
49.00	0.0010		
49.10	0.0009	0.0001	10%
49.90	$\simeq 0.0001(10^{-4})$		
50.00	1.0×10^{-7}	$\simeq 0.0001$	100%
51.00	1.01×10^{-11}		
51.10	9.19×10^{-12}	$\simeq 0.09 \times 10^{-11}$	8.9%
60.00	1.10×10^{-12}		
60.10	1.09×10^{-12}	0.01×10^{-12}	0.91%
70.00	6.00×10^{-13}		
70.10	5.98×10^{-13}	0.02×10^{-13}	0.33%

*50.00 ml of 0.1000 *M* HCl titrated with 0.1000 *M* NaOH.

coated with AgCl dipping into a fixed concentration of HCl, usually about 0.10 *M*. This solution is separated from the test solution by a glass membrane from 0.03 to 0.10 mm thick, which allows for electrical contact between the inner (reference) and outer (unknown) solutions. As the H^+ activity changes in the outer solution, there is a change in potential across the glass membrane and a corresponding change in the potential of the entire reference-indicator pair. Thus, the potential will change in a manner similar to that discussed above, the titrant being either H^+ or OH^-.

For redox reactions such as Fe^{2+} reactions with $Cr_2O_7{}^{2-}$, one can simply use a platinum wire as the indicator electrode. It monitors the potential change of the Fe^{2+},Fe^{3+} couple and the $Cr_2O_7{}^{2-}$,Cr^{3+} system.

The equivalence point is detected by any of several methods. The easiest is to measure the total volume of titrant delivered when the midpoint of the steep portion of the titration curve is reached. This can result in some error, however, since not all titrations yield a perfectly symmetrical curve. In most instances, the error introduced in using the midpoint is small and, therefore, acceptable. Alternately, one may choose to follow either the first or second derivative of the change in potential versus volume of titrant added. The second-derivative value is zero at the equivalence point, while for the first derivative, the equivalence point is taken at the maximum of the single peak generated. These derivatives may be obtained manually, but are more easily obtained with simple analog circuits. Of course, with a digital computer tied in, the derivatives can be calculated in real time, and the computer can stop the titration very shortly beyond the equivalence point.

For precipitation titrations, an indicator electrode which is sensitive to one of the ions forming the precipitate is used. For example, in halide titrations with $AgNO_3$, a silver wire electrode is used since it is sensitive to Ag^+ activity.

A useful experiment is one in which a potentiometric titration is done under computer control. A block diagram of the experimental setup is shown in Fig. 14-2.

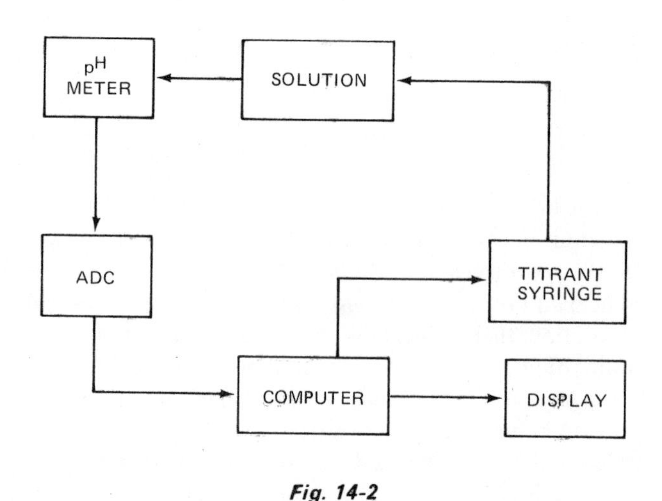

Fig. 14-2

Block diagram of the computer-controlled potentiometric titration experiment. [From *Chem. Tech.* **2**, 681 (1972), reprinted with permission of the American Chemical Society, copyright holder.]

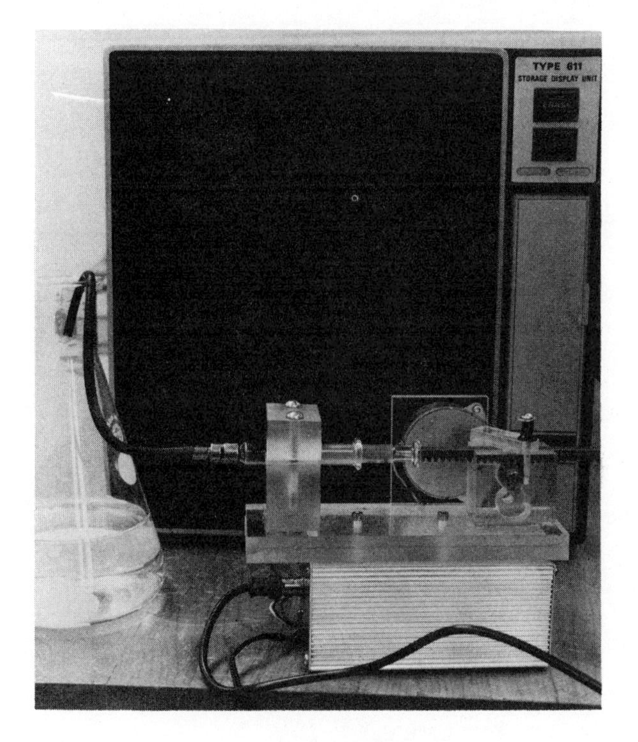

Fig. 14-3

Computer-driven syringe apparatus (stepper motor visible directly behind the syringe plunger). The cathode ray tube display unit (also under computer control) is also shown. [From *Chem. Tech.* **2**, 681 (1972), reprinted with permission of the American Chemical Society, copyright holder.]

Here a simple digital stepping motor is used to control titrant delivery. It advances a step each time a pulse is received from the computer and, via a rack and pinion gear, pushes a syringe plunger forward to deliver titrant (Fig. 14-3). The circuit for pulse control of the stepping motor is given in Appendix E.

The pH of the solution can be monitored and either raw or transformed data (as in a differential titration) can be displayed on the oscilloscope shown. Such niceties as slowing the rate of titrant delivery as the endpoint is reached are also possible. Curves may also be plotted on a Teletype or a simple X–Y plotter (Fig. 14-4). Again, titrant addition can be controlled to maintain a null point either for control or kinetic studies.

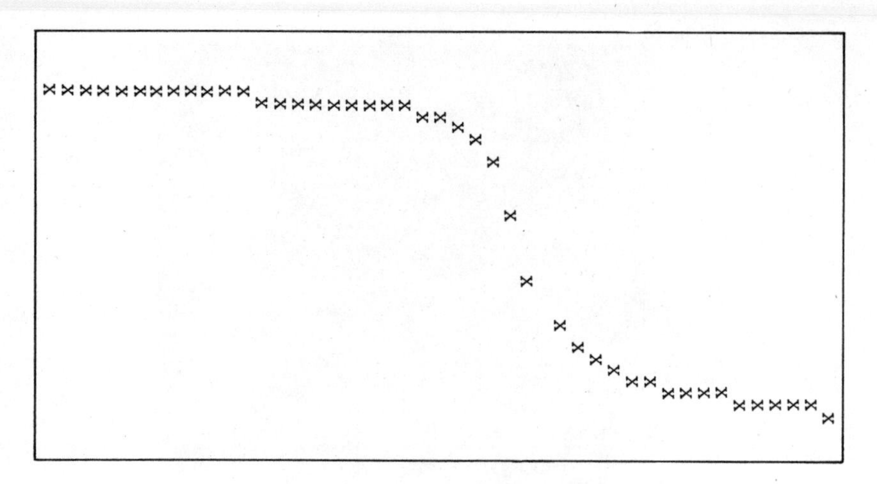

Fig. 14-4a
Teletype plot of the titration of a strong base with HCl.

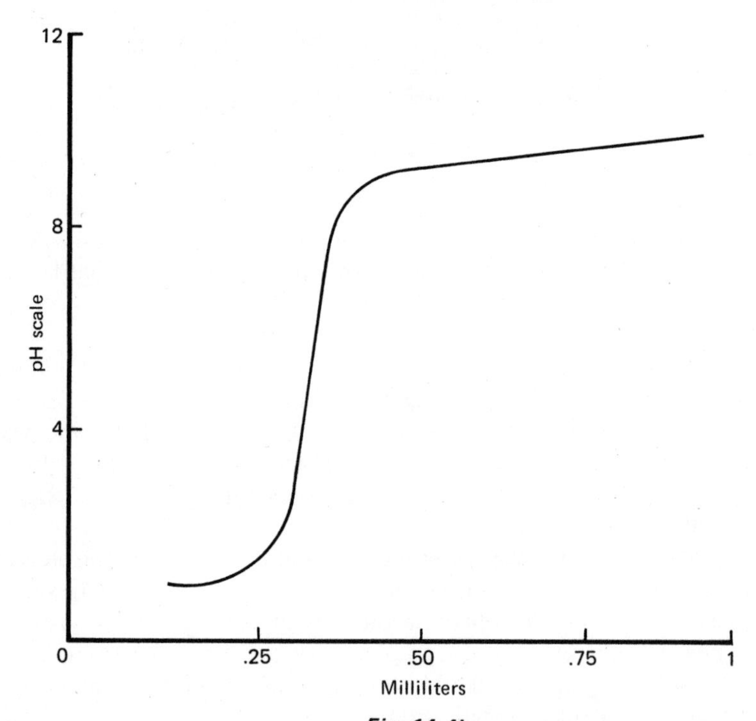

Fig. 14-4b
X–Y plot of the titration of a strong acid with NaOH.

EXPERIMENTAL

Equipment Needed

1 EL Digi-designer
1 $\pm$15-V power supply
3 Operational amplifiers (SN 72741)
2 Variable resistance boxes
2 1-MΩ 5% resistors
6 Capacitors: two 0.1 μF, two 0.22 μF, and two 0.47 μF
1 Oscilloscope
1 pH meter with terminals provided for output of the measured voltage
1 Magnetic stirrer and stirring bar
1 Stepper motor syringe driver (see Appendix E)
1 Laboratory computer system with digital and analog I/O interfaces (including one for an X–Y plotter or storage display oscilloscope)

Chemicals

Obtain samples of weak and strong acids and bases (e.g., acetic acid, HCl, NaOH, etc.). The titrants should be approximately 1.0 *M*, while the unknown should be 0.01 to 0.10 *M*.

Procedure

A. Potentiometric Tritration without Signal Conditioning

1. Attach the pH meter output directly to the input channel of the analog-to-digital converter.
2. Assemble the titration apparatus as shown in Fig. 14-2.
3. Using the appropriate gain setting and timing parameters, write a BASIC program which will measure the solution potential at sufficiently short intervals, relative to the rate of titrant addition, to develop an adequate titration curve. Include in your program statements which will cause the curve to be plotted on an appropriate display device. Determine the end point by a suitable technique (e.g., the derivative method; a flow chart of such a program is shown in Fig. 14-5). Your instructor may also require you to calibrate the volume delivered in each step, and the correspondence between pH and measured voltage.
4. If, at the gain setting chosen, noise is too great, write a signal smoothing program to minimize this source of error.

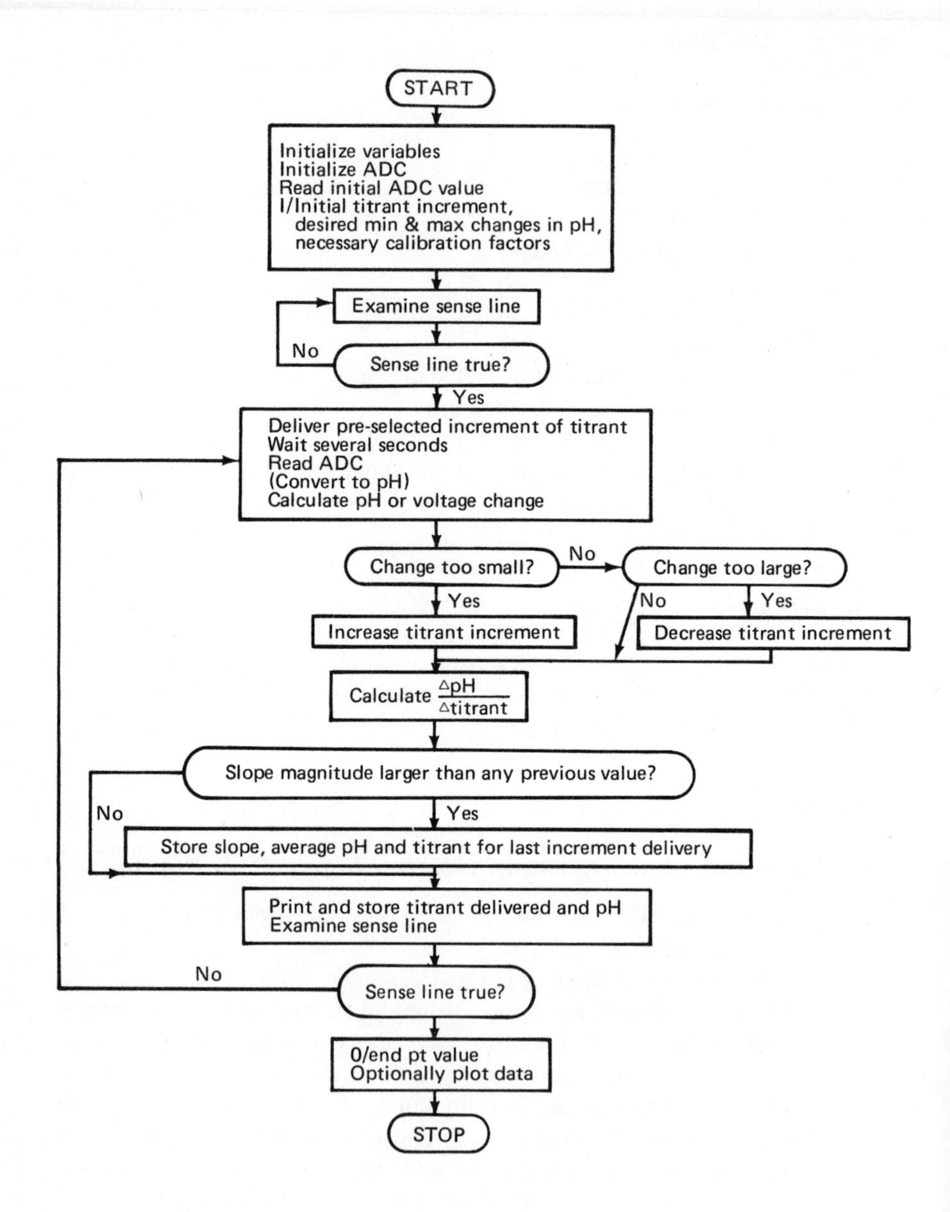

Fig. 14-5
Flow chart for titration and end-point determination.

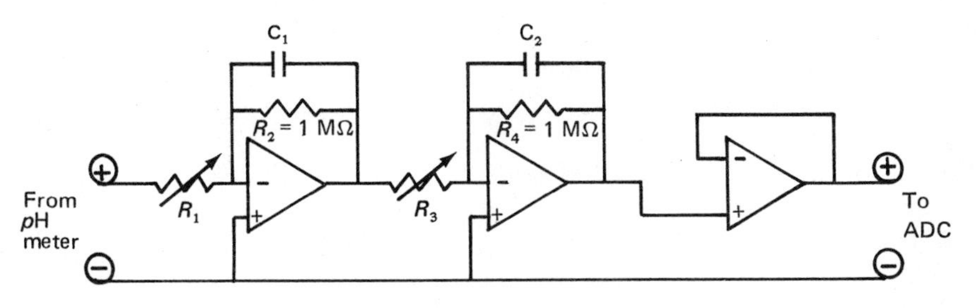

Fig. 14-6
A variable-gain operational-amplifier circuit.

B. *Potentiometric Titration Using Linear Amplifiers (Optional)*

1. Construct and power the linear amplifier circuit shown in Fig. 14-6, using 0.1-μF capacitors for C_1 and C_2. (The final amplifier on the right acts as a voltage follower, and is optional. Check with your instructor to find out whether it is needed on your computer system.) Adjust the variable resistors so that the amplified output is close to, but less than, the ADC $\pm$ maximum, at pH 0 and 14.
2. Observe the amplified output on the oscilloscope. Replace C_1 and C_2 with 0.22 μF capacitors and note any changes in the output voltage. Repeat the process with 0.47-μF capacitors. Now completely remove capacitors C_1 and C_2 from the circuit and note the results. What purpose do C_1 and C_2 serve?
3. Based on your results in step 2, choose a value for C_1 and C_2, and reinsert these capacitors into the circuit.
4. Using the pH meter as a source, measure and plot output voltage versus pH to verify that the amplifier is linear.
5. Now place the linear amplifier between the pH meter output and the ADC input. Using the program written for Part A, perform a titration. (Note that the ADC gain *must* be changed from that used previously to accommodate the output range of the amplifier.)
6. Repeat the experiment using software smoothing.
7. Compare these results to those obtained in Part A.

Experiment 15

On-Line Digital Computer Applications in Gas Chromatography

GAS CHROMATOGRAPHY (G.C.)

Gas chromatography is an analytical technique to achieve quantitative separation of mixtures. It is applicable to any samples which can be vaporized and introduced to the separation process in a gaseous state. A schematic diagram of the chromatographic instrumentation is shown in Fig. 15-1. More detailed discussion of G.C. theory, instrumentation, and procedures are given in refs. E-1 and E-2.

Quantitative Measurements in G.C.

Figure 15-2 illustrates a typical chromatographic output—or "chromatogram"—for a multicomponent mixture where each component has been well separated. Quantitative data are obtained by determining the area for each peak, A_i. If the detector *response* factor is identical for each component, the percent composition for the ith component is given by $(A_i/\Delta A_i) \times 100\%$. The area of a peak can be determined *manually* from the chart paper record of the output by triangulation, counting squares, or planimeter measurement.

Qualitative Measurements in G.C.

The precise elution time corresponding to the peak maximum (*retention time*) can be used for identification of each component in the sample. This feature depends on having retention-time data available for pure standards run under identical conditions of column material and size, length, temperature, and flow rate.

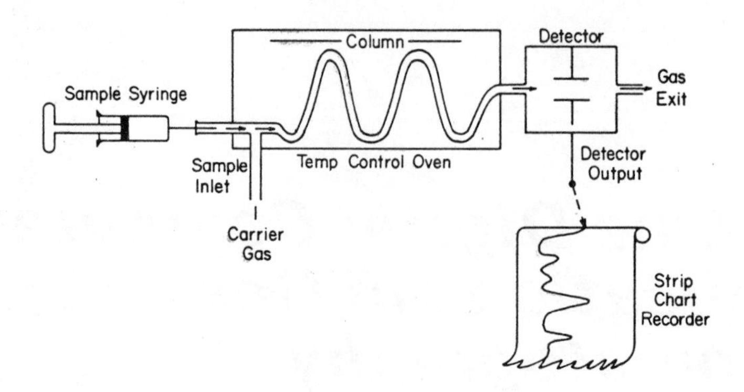

Fig. 15-1
Block diagram of G.C. instrumentation. [From *J. Chem. Educ.*, **48**, 438 (1971).]

COMPUTER PROCESSING OF CHROMATOGRAPHIC DATA

Assuming that chromatographic data can be tabulated and entered into the computer's memory, it becomes possible to use a computer program to process the data. Not only can the simple manual methods described above

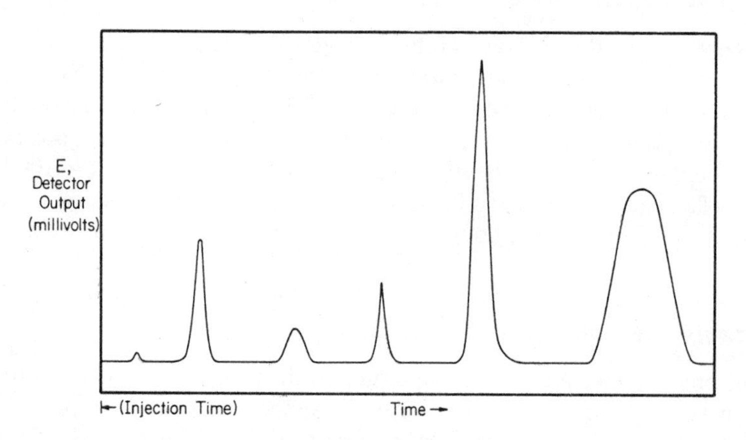

Fig. 15-2
Typical chromatogram for well-separated components.
[From *J. Chem. Educ.*, **48**, 438 (1971).]

be replaced, but processing problems which are too complex for manual methods can be considered and handled. We will defer until later the discussion of how chromatographic data are to be transferred to computer memory. First, let us consider how the data can be processed once they are acquired by the computer.

Peak Integration

The peak area is directly proportional to the *integral* of the peak waveform. Figure 15-3 and Eq. (15-1) show this relationship.

$$A \propto Q = \int_{t_1}^{t_2} (y - b)\, dt \tag{15-1}$$

Here, Q is the value of the integral, t is the time, y is the value of the G.C. output as a function of time, and b is the base-line value (which may or may not be time-dependent). Computer integration of data is a very simple operation, but one must consider first what information is available in memory, and how it can be used.

The G.C. data could be tabulated, for example, such that output values are measured at 1-sec intervals along the entire length of the curve and saved in memory. Assuming a total of 200 data points, the result would be an array where $Y(1)$ corresponds to the first datum in the array and the first time interval ($t = 1$ sec); $Y(100)$ corresponds to the 100th datum in the array and a time of 100 sec; and so on. This translation is shown in Fig. 15-4. The

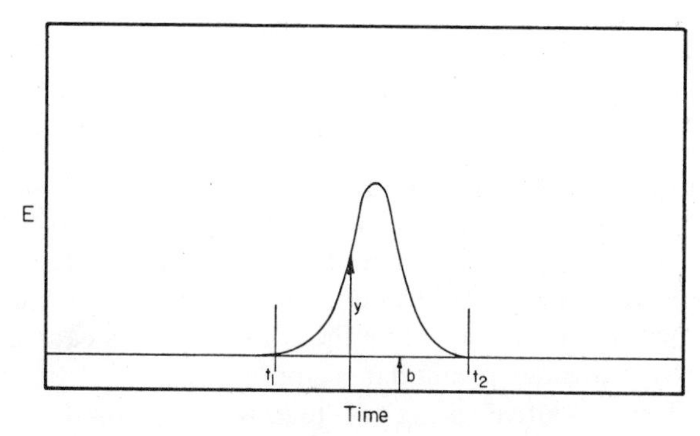

Fig. 15-3

Integration limits and variables for peak-area determination.
[From *J. Chem. Educ.*, **48**, 438 (1971).]

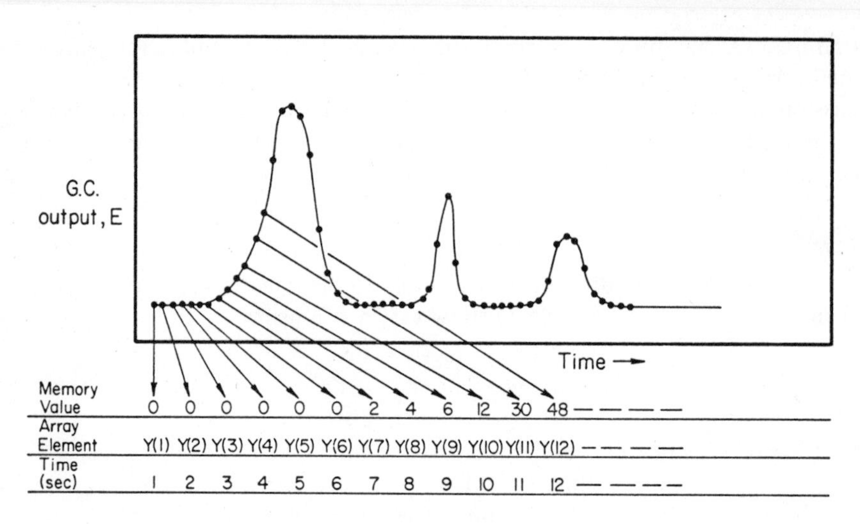

Fig. 15-4

Correspondence between experimental output and tabulated array in memory. [From *J. Chem. Educ.*, **48**, 438 (1971).]

computer program can access any element in the array (any data point) by specification of the proper subscript. Moreover, according to the storage scheme of Fig. 15-4, the subscript also corresponds to the time of the data point.

To generate a program to integrate peak areas, use the fact that an integral is defined by a summation of $y(t)\,\Delta t$ as Δt approaches zero:

$$Q \simeq \sum_{i=1}^{n} y_i\,\Delta t_i$$

For constant Δt,

$$Q \simeq \Delta t \sum_{i=1}^{n} y_i$$

The program might have the following features: assuming a zero base line, examine each data point consecutively; when data are consistently above some arbitrary threshold, assume the data are on the peak; add consecutive data points together until the data are consistently below the threshold again; the sum generated, when multiplied by the constant, Δt, is equivalent to the peak integral. (Note that a constant data sampling interval must be assumed. Also note that if only relative values of integrals are required—as is usually the case—it is not necessary to multiply the sum by Δt.) The *thresholding* algorithm described above is illustrated in the programming flow chart of Fig. 15-5.

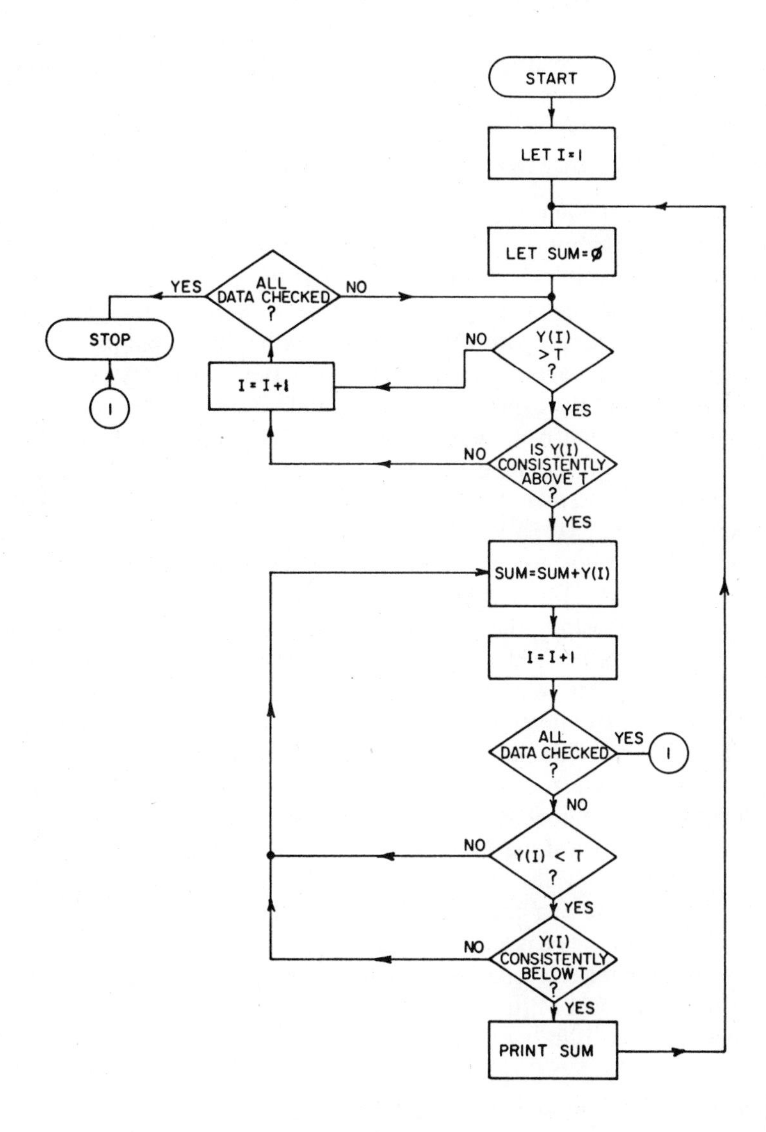

Fig. 15-5

Flow chart for program to integrate area under each peak in a chromatogram and print each result based on simple thresholding algorithm. $Y(I)$ is magnitude of datum and T is the threshold value. [From *J. Chem. Educ.*, **48**, 438 (1971).]

OTHER PROCESSING OBJECTIVES

In addition to the integration of individual peaks, several other processing functions may be required for gas-chromatographic data. Some of these will be outlined briefly here. Fewer details regarding the processing algorithms are presented, as these will be left to the student's discretion.

Precise Peak Location and Identification

The establishment of the time associated with the peak maximum (retention time) is important for identifying the component giving rise to the peak. The rigor with which one attempts to measure the retention time depends on several factors: (1) the accuracy with which the time base is known, i.e., the accuracy with which "zero" time has been established and the accuracy of the data-acquisition timing; (2) the reproducibility of the timing characteristics and the experimental data; and (3) the accuracy required for distinguishing qualitatively between possible components of the mixture.

Obviously, if the inherent accuracy and/or reproducibility of the experiment are poor, then it is foolish to develop a rigorously accurate peak-location program. Moreover, even if experimental accuracy and reproducibility allow rigorous programming, the required accuracy may be so undemanding that rigorous peak location processing is not warranted. One approach to establishing the retention time for each peak in this case would be to assume symmetrical peaks and compute the midpoint between the points where the data went above and below the threshold.

Should rigorous determination of retention times be both required and experimentally possible, alternative processing algorithms should be considered. For example, the peak data could be differentiated and the point at which the derivative goes through zero equated to the peak maximum. This is not a straightforward matter however, because the data in the computer's memory are discontinuous. Thus, mathematical differentiation can be achieved either by taking finite differences, or by fitting a quadratic or higher-order equation to the peak maximum and differentiating the equation. If the data are not completely *smooth*, the second alternative is preferred. (These algorithms and curve-fitting procedures are discussed in detail in the references listed in Bibliography Section D.)

Nonuniform Detector Response

To determine percent composition from peak areas, one must either be able to assume uniform detector response to all components, as described earlier, or else the individual response factors must be known. Assuming standard runs with each of the possible mixture components can be obtained,

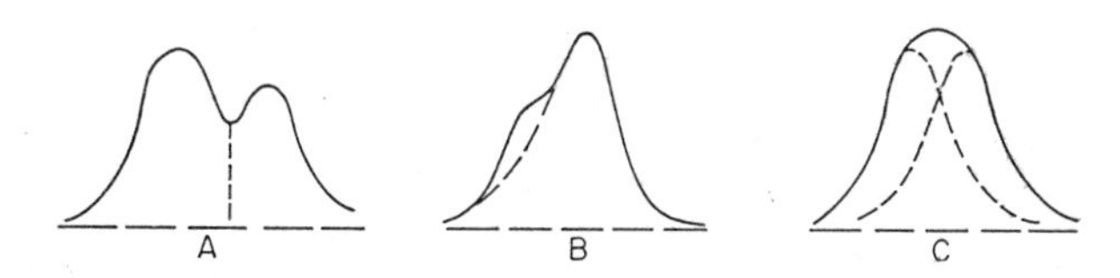

Fig. 15-6
Examples of unresolved G.C. peaks. (a) Two identifiable maxima; (b) one shoulder, one maximum; (c) two peaks fused to give one maximum. [From *J. Chem. Educ.*, **48**, 438 (1971).]

the response factors can be tabulated. Mathematical implementation of these factors to calculate correct percent compositions from peak areas should be obvious. The program development is left to the student.

Resolution of Overlapping Peaks

Frequently, gas chromatographic separations are incomplete because the optimum experimental conditions are unknown or unattainable. In such cases some chromatographic peaks may overlap as shown in Fig. 15-6.

Algorithms for resolving overlapping G.C. peaks have been discussed extensively in the literature (refs. E-3 to E-5). Many of these utilize the first and second derivatives of the G.C. data to recognize peak characteristics. These relationships are summarized in Fig. 6 of ref. E-3. It is left to the student to develop algorithms for G.C. data processing based on these characteristics. Perusal of the literature references given is strongly suggested.

Nonzero Base Lines

Up to this point we have assumed a base-line or background level which is zero or constant. However, it is not unusual to observe large base-line drifts in chromatographic experiments, as shown in Fig. 15-7. Problems of this nature are particularly prevalent when temperature programming is employed, as "column-bleed" may occur at elevated temperatures and contribute to the background level.

The data-processing algorithm to handle background drift could be similar to that designed to handle shoulders on large peaks. Estimating the base line for a given peak could be as simple as projecting a linear segment tangent to adjacent minima in the curves, or as complex as projecting a curved segment fit to the shape of the minima on either side of a peak. The experimenter must decide on how necessary an exotic processing program

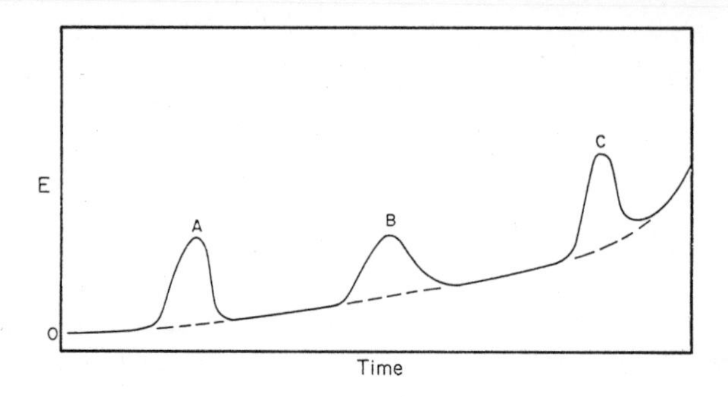

Fig. 15-7
Example of changing chromatographic base line. [From *J. Chem. Educ.*, **48**, 438 (1971).]

might be. This is generally based on the frequency with which difficult base-line problems are encountered in the systems studied. Generally, it is more practical to put effort into improving the experimental systems to avoid difficult processing situations, and utilize relatively straightforward processing algorithms.

Accurate Area Determinations

The thresholding algorithm suggested above as a method for peak-area measurements is obviously inaccurate. This is particularly so when applied to very small or very broad peaks. For large and/or sharp peaks the errors can usually be neglected.

In cases where accurate peak-area determinations are required, more rigorous processing algorithms must be used. More critical estimates of the points where the data begin to deviate from the background, and subsequently return to the background level, are required. The best approach probably is to look at the first and second derivatives, equating the start of the peak to the point where the second derivative changes from a consistent low or zero value to a definite positive value. The end of the peak should see the second derivative approaching a consistent low or zero value from the positive direction.

Noisy Data

Experimental data are rarely free from random fluctuations and background electronic noise. Because of this fact, the straightforward application of any processing algorithm which assumes "smooth" data curves may lead to serious misinterpretation of the data or complete failure. The magnitude of the problem is inversely related to the signal-to-noise ratio (S/N) of the data. If S/N never gets below 100:1, the data-processing program that assumes "smooth" data will probably work. When S/N falls somewhat below this value, one can no longer ignore the noise in the data-processing program.

The first step in solving the problem of processing noisy data is to attempt to isolate the instrumental source of the noise and minimize it. If the experimenter has implemented all reasonable means to minimize instrumental-noise contributions and the data still contain significant noise background, the only choice is to employ mathematical smoothing techniques.

In this approach a program is written which first processes the raw data to obtain a new smooth curve which hopefully fits the fundamental data. There are various mathematical approaches available. The student is referred to the article by Savitzky and Golay (D-9) and to Experiment 8 which provide a detailed discussion of smoothing methods and details for a least-squares technique to obtain smooth zero, first, and higher derivatives from raw noisy data. Once the smoothed data are obtained, the data-processing programs for extraction of chemical information may be applied.

Qualitative Identification

Computer identification of mixture components can be accomplished by establishing a table of standard retention times. A program to compare experimentally observed retention times with the tabular values can then achieve qualitative identification. There are two significant points to be kept in mind for this: (1) The retention-time-standard data must be obtained under

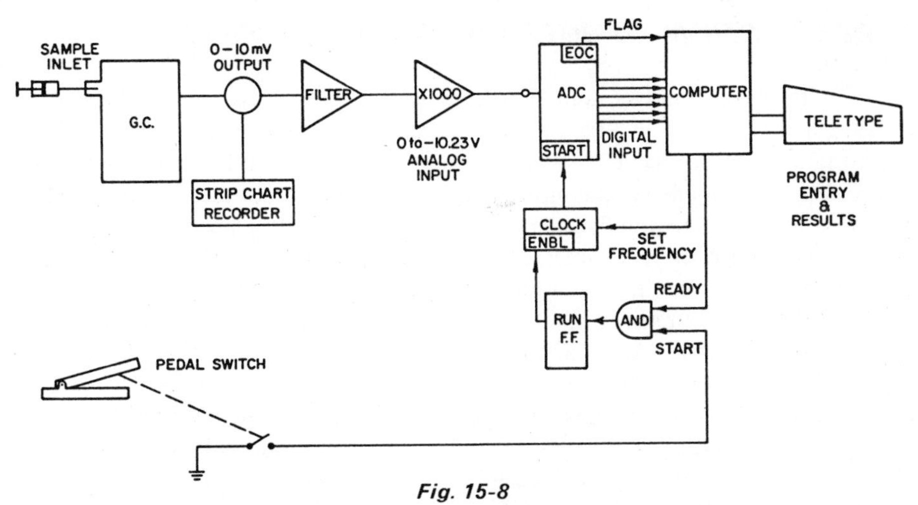

Fig. 15-8

Block diagram of instrumentation and interfacing for on-line computer system for G.C. experiment. [From *J. Chem. Educ.*, **48**, 438 (1971).]

identical experimental conditions to those for the unknown data; and (2) the comparison of observed retention times with the tabulated standards must allow for some experimental uncertainty in the measurements. A common approach is to assign some identification (I.D.) "window" to each standard retention time, such that any experimentally observed retention time that falls within that window is associated with that standard. The size of the window selected is a function of experimental reproducibility, the magnitude of the retention time, and the anticipated composition of mixtures. For example, a window of ± 10 sec might be assigned to a retention time at 100 sec when the next closest expected component has a retention time of 200 sec. However, if the next component has a retention time of 115 sec, it is necessary to establish a narrower I.D. window (e.g., ± 5 sec) if experimental reproducibility allows it.

ON-LINE COMPUTER PROGRAMMING AND INSTRUMENTATION

A block diagram of typical instrumentation and interfacing required for on-line execution of the G.C. experiment is shown in Fig. 15-8. All the interface components are provided in the laboratory. It is up to the student to make the appropriate connections, to select the proper conditions for operation, and to verify that the interface is working properly.

DESCRIPTION OF INTERFACE COMPONENTS

The basic interface and data-acquisition characteristics for a system such as that shown in Fig. 15-8 are described in detail in the introductory paragraphs in Sec. II. The student should be familiar with this description. For this experiment, special interface components include a voltage amplifier and filter, and a manually operated foot-pedal switch or push button, or a microswitch on the G.C. inlet. Figure 15-9 shows a circuit diagram for the amplifier/filter circuit. The input terminal should receive the direct G.C. output, and the amplifier–filter output should be connected to the analog-to-digital converter (ADC) analog input terminal.

The inlet switch, foot switch, or push button is also provided in the laboratory. It is used to synchronize the sample injection with the start of the automatic data-acquisition process. The electrical input and output of the switch can be accessed on the general-purpose interface. The switch output can be used to condition the AND gate input to enable the CLOCK at the start

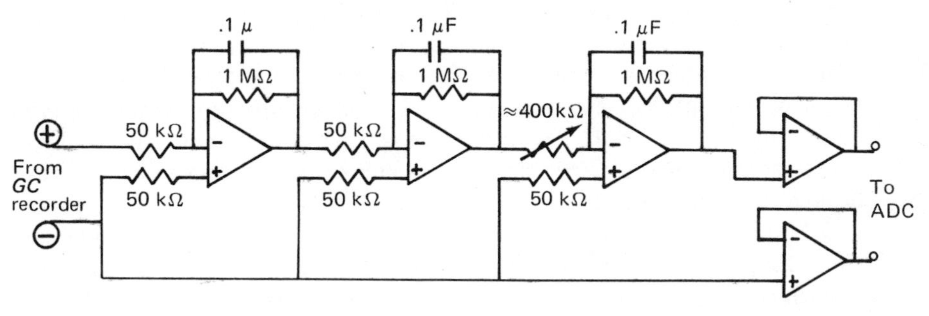

Fig. 15-9
Circuit diagram for a voltage amplifier and filter to be used in the G.C. experiment.

of the experiment. (The other AND gate input must have been set by the computer when it was ready to accept experimental data.) The data-acquisition rate is determined by the CLOCK frequency. That frequency can be set by the experimenter's program.

ON-LINE PROGRAMMING FEATURES

The experimenter's program must include the following considerations: (1) selection of data-acquisition rate, (2) synchronization with the start of the experiment, (3) data storage, and (4) "real-time" and "non-real-time" data processing.

Selection of the data-acquisition rate must be based on some knowledge of the desired data density. A rate which provides about 20 or 30 data points on a peak is adequate.

Data storage can be a serious problem. The computer has finite memory space available. Moreover, in most minicomputer BASIC configurations, data storage is available only in 255-word blocks. Thus, if one were to store all data points from an experiment, the total number could not exceed 255 unless the experimenter made provision to link blocks during the data-acquisition process. Moreover, a maximum of only 500 to 1000 data points could be stored—assuming minimal program space. Thus, for a typical G.C. experiment that runs several minutes, with a required data-acquisition rate of 10 pts/sec, there would not be enough storage space available.

At this point, the experimenter must consider "real-time" data processing, i.e., processing the data as they are being acquired. One advantage of this type of processing is that not every data point need be stored. For example, the simplest approach might involve looking at data as they come in, deciding whether or not the data are on a peak, saving only those data corresponding to peaks, and saving the corresponding times for start and finish of each peak. Using this approach, a chromatogram containing 10 peaks might require only 200 or 300 storage spaces in memory. When data acquisition is completed, the program can fetch the stored data and process as desired in "non-real-time."

Many other "real-time" data-processing functions might be considered. For example, the data might be integrated as they are taken. Thus, at the end of each peak, the peak integral would be available. The limitation to "real-time" programming is the amount of time available between data points and the time required for program execution between data points. You can use trial and error to determine how many program statements can be executed in a given time interval, or with the help of an oscilloscope, the computer, and the general interface, you can determine the approximate time per BASIC statement experimentally. A useful guideline for typical BASIC compilers supplied with common minicomputers is that each program statement will require between 2 and 10 msec of execution time.

EXPERIMENTAL

Equipment Needed

- 1 EL Digi-designer
- 1 ±15-V power supply
- 5 Operational amplifiers (SN 72741)
- 1 Variable resistance box
- 8 Resistors: five 50 kΩ and three 1 MΩ
- 3 Capacitors: 0.1 μF
- 1 Laboratory computer system with programmable clock and digital and analog I/O interfaces
- 1 Gas Chromatograph-Aerograph 202 with one 10-ft 20% SF-96 column and one 6-ft 20% Carbowax 30-mesh column, and a 1-mV recorder (or similar device)
- 1 Microswitch for G.C. inlet, or foot-pedal or push-button switch

Chemicals

1. For three well defined peaks, zero base-line: mixture of *n*-heptane, *n*-octane, and *n*-nonane*

2. For base-line drift w/o temp. programming†:

Fair results	0.5 ml pentane	and 10 ml MeOH	time 1.5 min
Good results	0.5 ml hexane		time 2 min
V. good results	0.5 ml heptane		time 6 min

3. For fused peaks:
   Fair—mixture of pentane, hexane, and heptane*
   Good—mixture of MeOH, pentane, and hexane†

*6-ft Carbowax 30-m column, 0.5 ml/sec flow rate. Column, detector, and injector temperatures were, respectively, 79°C, 200°C, and 129°C. (These three temperatures were the same for all other samples.)
†10-ft 20% SF-96 column, 1 ml/sec flow rate.

Procedure

The following assignments are listed in order of progressively increasing difficulty. Each student should complete assignments A1 to A3. For advanced work, complete any of the additional assignments. Two to four 4-hour lab-session equivalents are required, depending on number of alternative studies pursued. Programs must be developed outside of normal lab time.

A. Assuming Well-Separated Peaks and Zero Base Lines for Assignments 1 to 5

1. Enter synthetic G.C. data from paper tapes supplied by teaching assistant in the lab. Process the data with a simple thresholding algorithm to determine peak areas and percent composition of each component. Determine retention times using t_r = midpoint of threshold segment.
2. Interface G.C. to the computer and execute the data-acquisition-program segment, simply printing out the acquired data block.
3. Take a complete data block on-line and process as in assignment 1.
4. Same as assignment 3, except provide for correction for detector response factors.
5. Take G.C. data on-line and use "real-time" data processing to integrate each peak as it is observed, and provide same processed data as in assignment 1, except don't save any data points permanently.

B. Without Assuming Well-Separated Peaks or Zero Base Lines

1. Use an appropriate on-line processing approach to handle overlapping peaks.

2. Use an appropriate on-line processing approach for handling nonzero and nonconstant base-line problems.
3. Use a mathematical smoothing approach to satisfactorily handle noisy data.
4. Use an appropriate processing approach for obtaining very *accurate* peak-area and retention-time measurements.

Experiment 16

On-Line Digital Computer Applications for Kinetic Analysis

The experiment described here is one in which both kinetic methods of analysis and a computer employed in an on-line configuration are used to solve an analytical problem (see ref. E-6). In this experiment you will determine the concentration of glucose in the 10–100 ppm range by measuring its oxidation rate in the presence of the enzyme glucose oxidase. The reaction is

$$RCHO + O_2 + H_2O \underset{\text{oxidase}}{\overset{\text{Glucose}}{\rightleftharpoons}} RCOOH + H_2O_2 \qquad (16\text{-}1)$$

The hydrogen peroxide produced reacts rapidly with iodide ion to produce an equivalent amount of iodine (see refs. E-7, E-8).

$$H_2O_2 + 2I^- + 2H^+ \xrightarrow{\text{Mo(VI)}} I_2 + 2H_2O \qquad (16\text{-}2)$$

The use of a molybdate catalyst [Mo(VI)] ensures that reaction 2 is faster than the enzymatic reaction, so that the rate of iodine production is equal to the rate of glucose consumption. The overall reaction is

$$RCHO + O_2 + 2I^- + 2H^+ \xrightarrow[\text{Mo(VI)}]{\text{Glucose oxidase}} RCOOH + H_2O + I_2 \qquad (16\text{-}3)$$

The rate at which iodine is produced is given by

$$d[I_2]/dt = KG_t \qquad (16\text{-}4)$$

where G_t is the glucose concentration at time t and K is a pseudo-first-order rate constant. A rotating platinum electrode is used to detect the rate of iodine production by monitoring the electrolysis current i_t at the electrode. The electrolysis current is proportional to the iodine concentration at time t:

$$i_t = K'[I_2]_t \qquad (16\text{-}5)$$

where K' is dependent upon characteristics of the cell, electrode, and polarizing voltage. The change in current from time t_1 to t_2 is given by

$$\Delta i = K'\{[I_2]_2 - [I_2]_1\} = K'(\Delta[I_2]) \tag{16-6}$$

If the current interval is small enough to ensure that the total change in the glucose concentration will be small, the iodine concentration will change *linearly*. Thus, $d[I_2]/dt$ in Eq. (16-4) can be approximated by $\Delta[I_2]/\Delta t$. Solving Eq. (16-4) for $\Delta[I_2]$ and substituting this value into Eq. (16-6) gives

$$G_t = (\Delta i/KK')(1/\Delta t) \tag{16-7}$$

Substituting $G_t = G_0 - i_t/K'$ into Eq. (16-7), one obtains

$$G_0 = \frac{1}{K'}\left\{\frac{\Delta i}{K}\frac{1}{\Delta t} + i_t\right\} \tag{16-8}$$

Thus, Eq. (16-8) predicts that glucose concentration is a linear function of the reciprocal of the elapsed time (Δt) required to traverse a predetermined current interval Δi. It has previously been shown (see ref. E-9) that Eq. (16-8) is valid to within 1% if the measurement interval consumes less than 5% of the total glucose present when the measurement is started.

In the present experiment, a reliable quantitative analysis of glucose via a computer-assisted kinetic method is the aim. Accordingly, the programs written should provide the following functions: data acquisition from kinetic runs on standard samples; least-squares fitting of a linear calibration function; data acquisition from unknown samples; and print-out of analytical results. In addition, several options might be included in the on-line program, such as the possibility of averaging a series of runs before analytical computations are made, rejecting any given run, or determining a new calibration curve. The program should also allow for selection of the data-sampling and Δi interval.

A flow chart of such a kinetic-analysis program is provided in Fig. 16-1. Figure 16-2 is a more detailed flow chart for the data-acquisition segment (which is included in the boxes labeled "Run Expt." in Fig. 16-1).

As the data-acquisition process is outlined in Fig. 16-2, that part of the program begins with an initialization of a variable T (proportional to time) and then starts a clock or timer. Digitized data is then acquired at time intervals determined by the clock frequency (if hardware control is used) or by the program (if software control is used). At the beginning, each digitized datum is examined and compared with an initial threshold. As long as the incoming data are not less than this threshold value, no further comparisons are made. When that does happen, the time interval between the taking of the datum which was less than the initial threshold and the datum which first becomes less than the final threshold is computed. This is possible because of the known data-acquisition rate. (Note that the directions of the

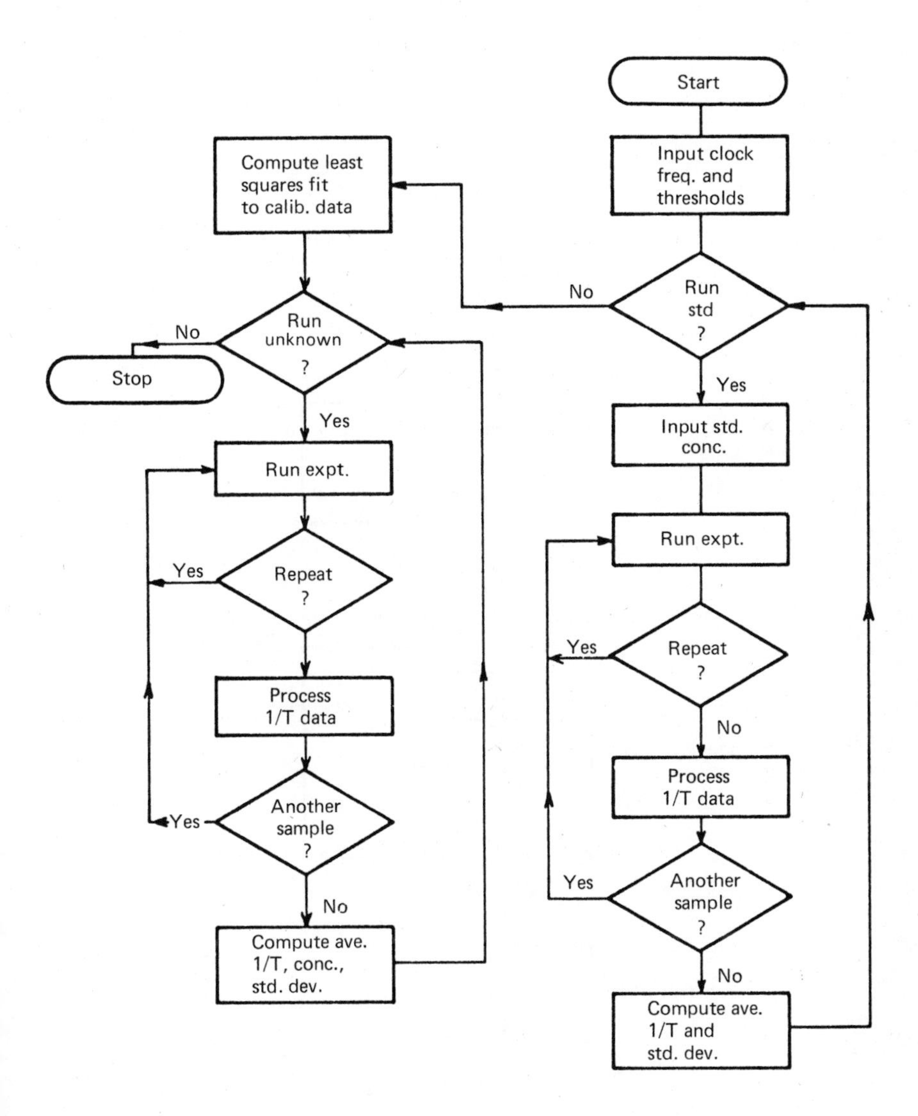

Fig. 16-1

Overall flow chart for kinetic-analysis experiment. [From *J. Chem. Educ.*, **49**, 717 (1972).]

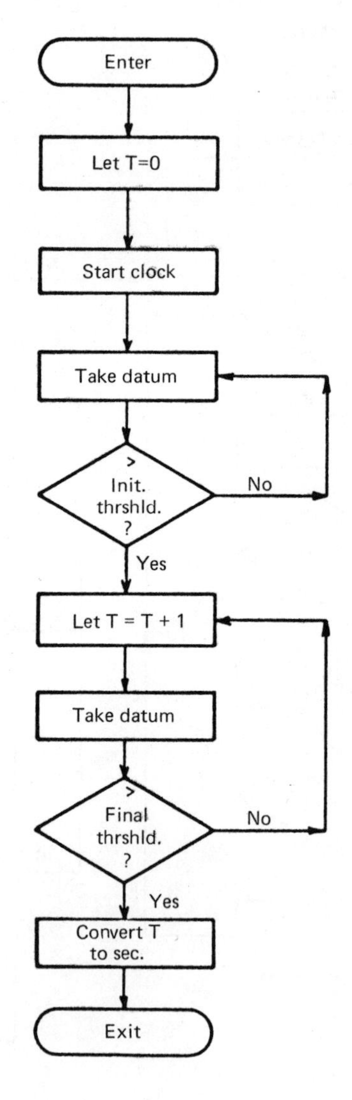

Fig. 16-2

Flow chart for data acquisition only. [From *J. Chem. Educ.*, **49**, 717 (1972).]

inequality signs in threshold comparison boxes reflect the fact that the *digitized* values *decrease* during the experiment. If the digital values were *increasing* with time, the inequality signs should be reversed.) Because the noise level of the output signal is usually $<1\%$, there was no need to incorporate noise-rejection steps in the data-acquisition program. If noise were a problem, an averaging algorithm could be used when monitoring the data.

A typical computer/operator dialogue and results print-out is shown in Fig. 16-3. Representative analytical results over the range 10 to 100 ppm

```
RUN
THIS PROGRAM ANALYZES THE KINETICS OF THE OXIDATION OF GLUCOSE BY
THE ENZYME GLUCOSE OXIDASE.

FOR ALL YES/NO QUESTIONS WHICH THE USER OF THIS PROGRAM IS ASKED,
TYPE 1 FOR YES AND Ø FOR NO. FOR ALL OTHER QUESTIONS, TYPE IN THE
NUMERICAL VALUE OF THE QUANTITY BEING SOUGHT.

WHAT ARE THE INITIAL AND FINAL THRESHOLDS FOR ALL RUNS?7ØØ,2ØØ

DO YOU WANT TO RUN A STANDARD?1

READY TO BEGIN EXPERIMENT
CONC. = 1ØØ      TRIAL 1       T = 6.7      1/T = Ø.149254
DO YOU WANT TO REPEAT THIS TRIAL?Ø
ANOTHER TRIAL ?1

READY TO BEGIN EXPERIMENT
CONC. = 5Ø       TRIAL 3       T = 14.4     1/T = 6.94444E-Ø2
DO YOU WANT TO REPEAT THIS TRIAL?Ø
ANOTHER TRIAL?Ø
CONC. = 5Ø       AVERAGE 1/T FOR 3          TRIALS = 6.88124E-Ø2
     STD. DEVIATION = 7.2Ø887E-Ø4

DO YOU WANT TO RUN A STANDARD?Ø
LEAST SQUARES FITTED LINE IS —
     1/T = 1.44942E-Ø3      C + -4.38669E-Ø3
     1/T IN SEC AND C IN PPM

CONC.       AVE. 1/T                 DEVIATION
1ØØ         Ø.14Ø376                 1.78814E-Ø4   (Ø.13%)
 5Ø         6.88124E-Ø2             -7.28279E-Ø4   (1.1%)
 25         3.Ø9792E-Ø2              8.69516E-Ø4   (2.8%)
 1Ø         1.Ø4274E-Ø2             -3.19956E-Ø4   (3.1%)

DO YOU WANT TO RUN AN UNKNOWN?Ø
RECORD VALUES IN NOTEBOOK. GOODBYE.
```

Fig. 16-3

Typical computer/operator dialogue for kinetic glucose determination.

glucose are also given in Fig. 16-3. A typical calibration plot for this system is given in ref. E-9. A distinct nonzero intercept is normally observed, fixing the lower limit of sensitivity to 10 ppm.

The reaction is carried out in a small cell having a total capacity of about 5 ml. The cell is immersed in a thermostated liquid bath in order to facilitate temperature control. (The details of cell dimensions and apparatus can be found in Appendix E.)

The electronic circuitry used in the experiment is depicted in Fig. 16-4. This circuitry performs the function of controlling the potential of the monitoring electrode, measuring the electrolysis current, and making the analog data acceptable for input into the computer. Since most common ADC's can readily monitor voltages between −10 and +10 V, it is convenient to convert the signal from the reaction cell to a voltage and amplify it. The circuit shown in Fig. 16-4 amplifies and converts the signal to a value between 0 and −10 V. Obviously, this is an arbitrary selection of amplification to generate a voltage convenient for certain ADC's. Some modification may be necessary if this is inconsistent with the range available on the ADC used.

The −15-V input to the summing point of amplifier A_1 in Fig. 16-4 is included to bias the output of A_1 to about +1.0 V. Because the gain of A_2 is −10, the initial output bias is about −10 V. When electrolysis current increases in the cell, the output of A_1 goes negative, while the output of A_2 goes positive. The diode in the feedback loop of A_2 prevents the output of A_2 from going more positive than 0 V. Thus, the output is kept within the acceptable input range of the ADC (0 to −10 V).

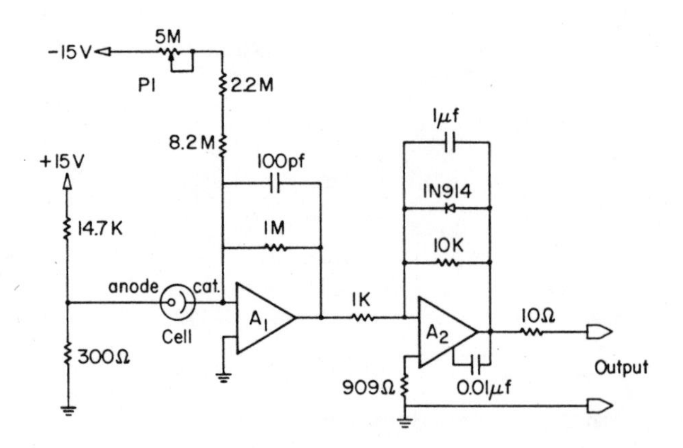

Fig. 16-4

Circuit for kinetic glucose determination experiment. [From *J. Chem. Educ.*, **49**, 717 (1972).]

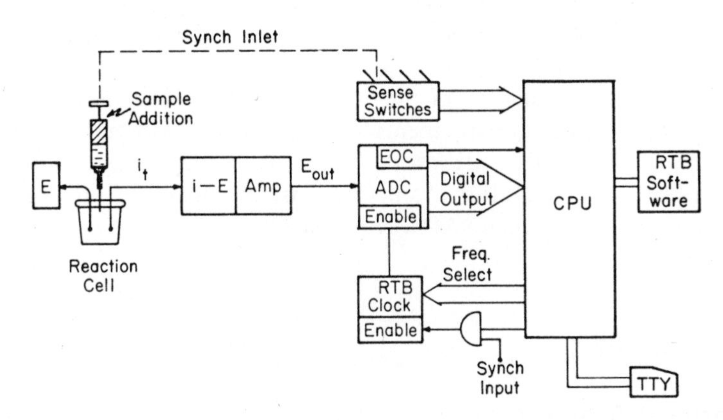

Fig. 16-5

Schematic diagram for on-line kinetic glucose determination. [From *J. Chem. Educ.*, **49**, 717 (1972).]

Potentiometer *P*1 should be adjusted at the beginning of each experiment, as the buffer and enzyme solutions may vary in residual current. However, if a series of experiments are being run where only analytical Δi values are measured, the bias voltage need not be adjusted for each run as long as the lower threshold for Δi is about 0.1 μA above the initial bias value.

The amplification factors for the circuitry of Fig. 16-4 provide a 10-V output swing for a 1-μA change in cell current (0.1 μA/V conversion factor). This represents the optimum condition for the analytical experiments described here and for the particular cell and electrode characteristics. The amplification factor can be lowered satisfactorily by decreasing the feedback resistance for A_2. This would be necessary for experiments to monitor total reaction times.

The schematic diagram for the overall experimental setup, including the on-line computer, is given in Fig. 16-5. The specific interfacing logic and connections made will depend on the particular experimental objectives. For example, if only a measurement of the Δt for traverse of a specified current interval after initiation of the reaction is required, no critical synchronization features need be incorporated in the interface. However, if it is desired to monitor the entire course of the reaction, starting at t_0, the computer must be made aware of the instant when sample is injected. This can be accomplished easily within the RTB system by utilizing manual sense switches or sense lines. An uncertainty of ± 0.5 sec in t_0 can be tolerated for this experiment, as reagent mixing is not instantaneous, and the reaction half-life is greater than 30 sec, even for the fastest conditions considered here. The Introduction to Section II discusses details of the general-purpose interfacing capabilities of a RTB system which allow synchronization of experimental and computer functions.

While the main purpose of the experiment is to introduce computer-assisted reaction-rate measurements (ref. E-10) and kinetic methods of analysis (ref. E-11), it also introduces the use of an enzyme as an analytical reagent (ref. E-12). Operational amplifiers are employed to condition the analog data to accommodate ADC requirements.

EXPERIMENTAL

Equipment Needed

- 1 EL Digi-designer
- 1 ±15-V power supply
- 2 Operational amplifiers (one SN 72741 and one Analog Devices P501A)
- 9 Resistors: 8.2 MΩ, 2.2 MΩ, 1 MΩ, 14.7 kΩ, 10 kΩ, 1 kΩ, 909 Ω, 300 Ω, and 10 Ω
- 1 Variable resistor (5 MΩ)
- 7 Capacitors: one 100 pF, five 0.01 μF, and one 1 μF
- 1 Diode (1N914)
- 1 Cell (see Appendix E)
- 1 Magnetic stirrer and stirring bar
- 1 2.5-ml syringe
- 1 Laboratory computer system with programmable clock and digital and analog I/O interfaces

Chemicals

All solutions used in the experiment should be prepared in deionized water and stored until use in a cold room at 5°C.

Buffer-Catalyst

The buffer-catalyst is prepared by dissolving 82 g of potassium dihydrogen phosphate, 42 g of potassium monohydrogen phosphate, and 13 g of ammonium molybdate $[(NH_4)_6Mo_7O_{24} \cdot 4H_2O]$ in deionized water and diluting the solution to a volume of 1 liter. This solution is stable indefinitely at 5°C.

Potassium Iodide

Dissolve 83 g of potassium iodide in deionized water and dilute to 1 liter. This solution is stable for several weeks when stored at 5°C.

Glucose Oxidase

Dissolve 0.06 g of glucose oxidase (Sigma Type II) in 50 ml of deionized water. This solution is stable for one to two weeks at 5°C.

Composite Reagent

Prepare 50 ml of composite reagent by mixing 20 ml each of both the buffer catalyst and potassium iodide solutions and 10 ml of the enzyme solution. This solution should be stored in a flask which is protected from the light.

Glucose Standards

Prepare standard glucose solutions of 10, 25, 50, and 100 ppm by the dilution of a 1000-ppm glucose solution prepared by dissolving 1.000 g of C.P. reagent sugar in deionized water and diluting to 1 liter. These standard solutions should be stored at 5°C.

Procedure

A. Quantitative Analysis

The detailed procedure presented here is for experiments where the relationship between G_0 and $1/\Delta t$ for a specified current interval Δi is determined and used for quantitative analysis. (The fundamental equations are provided in the introduction to this experiment.) Basically, the experiments involve establishing an interval, Δi, and programming the computer to monitor the output signal to determine Δt required to traverse the lower and upper signal thresholds bracketing Δi.

1. Wire the circuit shown in Fig. 16-4 making any necessary modifications to convert to appropriate voltage level for your laboratory computer. Use the P501A for A_1 and the 72741 for A_2. Wire 0.01 μF capacitors between the power supply leads and ground for each amplifier.
2. Write a program to determine the time required for the output signal to traverse the region between an initial and final threshold specified in the program (Δi). When the final threshold is passed, the program could cause printing of the trial number, the time, and the inverse time, and ask if another trial is to be attempted. Before

executing another run, the user should turn off the circuit power supply, turn off the stirrer, and then drain and rinse out the reaction cell. (*Note*: The initial adjustment of *P*1 could be checked before rerunning the experiment by programming the computer to sample and print the initial output bias voltage.) A 10-Hz clock rate gives sufficient accuracy when the averages of at least three trials for each standard or unknown are used. Different clock rates can be used to see if any increase in accuracy and precision is obtained. Refer to the flow charts of Figs. 16-1 and 16-2.

3. Next, adjust all reagents, standards, and samples to the working temperature by immersion in a water bath. Rinse the sample compartment with deionized water and use a siphon to remove the water from the compartment. Add 1 ml of composite reagent from the hypodermic syringe to the cell. Start the stirrer. Turn on the power supply to the monitoring circuit and wait about 5 to 10 sec until the initial voltage stabilizes. (*Note*: The electrodes should be polarized before use, or, alternatively, several trials can be performed before activating the computer. Otherwise, the initial voltage set by *P*1 in Fig. 16-4 will not remain stable for the first few trials.) Adjust *P*1 to bring the voltage to about −9.5 V. Initiate execution of the computer program. Add 1.00 ml of glucose sample to the cell. Use a manual sense switch to indicate to the computer when it should begin to monitor the output signal. Since a 1.0°C temperature rise causes a 10% rise in the reaction rate, the water-bath temperature should be kept to ±0.1°C to ensure ±1% accuracy.
4. Determine the concentration of an unknown solution.

B. Kinetic Studies—Determination of Rate Constants and Reaction Order

1. Modify the program used in Section A1 to calculate rate constants and reaction orders.
2. Run the program using several concentrations of glucose.
3. If need be, add a smoothing routine to improve the data.

C. Effects of Experimental Parameters (Optional)

1. Refer to the original literature (ref. E-7, E-8, and E-9) for a discussion of various features open to study.
2. Investigate the effects of experimental parameters on the accuracy and precision of quantitative data (e.g., Δi and clock rate, temperature, pH, and solvent effects).

Experiment 17

Computer-Controlled Colorimetry

It is often necessary to determine the composition of a metal complex which absorbs light in the visible region (ref. E-13). To this end, Job (ref. E-14) introduced a method of continuous variations in 1928 which was improved upon by Vosburgh and Cooper in 1941 (ref. E-15).

In this experiment Job's method (the method of continuous variations) is used to elucidate the composition of some complexes formed between Ni^{2+} and ethylenediamine (EN) in aqueous solution. Since Job's method requires many, many manipulations of samples and blanks, the experiment can become quite tedious. A spectrophotometer that is on-line to a small digital computer can be used to relieve the tedium and reduce the time required to perform this experiment, as well as to record the data and subsequently reduce and display it.

Consider a generalized chemical system in which a metal ion (M) reacts with any number of ligand ions (X) to produce a complex ion of some unknown molecular composition (MX_n). This reaction is given in Eq. (17-1):

$$M + nX \rightarrow MX_n \quad (17\text{-}1)$$

In order to determine n, the number of ligand molecules bound to the metal ion, Job proposed the following method. The concentration of the metal ion and ligand ion in a number of solutions is varied, but their sum is kept constant. Thus

$$[M]_0 + [X]_0 = C \quad (17\text{-}2)$$

where $[M]_0$ and $[X]_0$ are the concentrations of M and X before any reaction. If we define f as the mole fraction of ligand in the solute, then

$$[X]_0 = fC \quad (17\text{-}3)$$

and

$$[M]_0 = (1 - f)C \quad (17\text{-}4)$$

If the reaction shown in Eq. (17-1) produces only one product, MX_n, then the concentrations of free M and X are

$$[M] = (1 - f)C - [MX_n] \tag{17-5}$$

$$[X] = fC - n[MX_n] \tag{17-6}$$

Consider now the formation constant for Eq. (17-1):

$$K_n = [MX_n]/[M][X]^n \tag{17-7}$$

Solving this expression for $[MX_n]$, we get

$$[MX_n] = K_n[M][X]^n \tag{17-8}$$

Substituting Eqs. (17-5) and (17-6) into Eq. (17-8) gives

$$[MX_n] = K_n\{(1 - f)C - [MX_n]\}\{fC - n[MX_n]\}^n \tag{17-9}$$

Since K_n, n, and C are constants, $[MX_n]$ is simply a function of f, the mole fraction of ligand. The condition of interest is, of course, when $[MX_n]$ is at its maximum. At the maximum, the derivative of a function has the value zero. Thus, differentiating Eq. (17-9) with respect to f and setting $d[MX_n]/df$ equal to zero, we obtain the equation for n as a function of f at the maximum $[MX_n]$:

$$n = f/(1 - f) \tag{17-10}$$

Equation (17-10) says that the value of n can be calculated directly from the value of f at which $d[MX_n]/df$ is a maximum.

In practice, the most convenient and nondestructive method of measuring $[MX_n]$ is spectrophotometry. To determine $[MX_n]$ via spectrophotometry, Beer's law

$$A = abc \tag{17-11}$$

must be applied; here A is the absorbance of the solution, a is the molar absorptivity of the absorbing species, b is the breadth of the cell, and c is the concentration of the absorbing species, expressed in moles per liter. Thus, a knowledge of a and b and a measurement of A gives the concentration of the absorbing species directly.

Consider now a more complex chemical system in which the metal reacts with the ligand in a number of reactions:

$$\begin{aligned} M + X &\rightarrow MX \\ MX + X &\rightarrow MX_2 \\ MX_2 + X &\rightarrow MX_3 \\ MX_{n-1} + X &\rightarrow MX_n \end{aligned} \tag{17-12}$$

The overall reaction is identical to that given in Eq. (17-1). Typically, the only nonabsorbing species in this scheme is the ligand X. Thus, a measure of the absorbance of a solution of M with X will, in general, be a complex quantity consisting of the sum of the absorbances of all absorbing species present. We can write for the observed absorbance:

$$A_{\text{obs}} = A_{\text{M}} + \sum_n A_{\text{MX}_n} \tag{17-13}$$

or, by substitution from Beer's law,

$$A_{\text{obs}} = a_{\text{M}}[\text{M}] + \sum_n a_{\text{MX}_n}[\text{MX}_n] \tag{17-14}$$

where a cell breadth of 1 cm is assumed. Recalling Eq. (17-5), we can rewrite Eq. (17-14) to give

$$A_{\text{obs}} = a_{\text{M}}\{(1 - f)C - \sum_n [\text{MX}_n]\} + \sum_n a_{\text{MX}_n}[\text{MX}_n] \tag{17-15}$$

or

$$A_{\text{obs}} + (1 - f)Ca_{\text{M}} + \sum_n \{[\text{MX}_n](a_{\text{MX}_n} - a_{\text{M}})\} \tag{17-16}$$

Consider now the quantity Y, which is the difference between the observed absorbance of the mixture of M and X and the expected absorbance if there were no reaction; Y can be expressed as

$$Y = A_{\text{obs}} - a_{\text{M}}[\text{M}]_0 \tag{17-17}$$

substituting $[\text{M}]_0$ from Eq. (17-4), we get

$$Y = A_{\text{obs}} - (1 - f)Ca_{\text{M}} \tag{17-18}$$

Note that Eq. (17-16) can be rearranged to give

$$A_{\text{obs}} - (1 - f)Ca_{\text{M}} = \sum_n [\text{MX}_n](a_{\text{MX}_n} - a_{\text{M}}) \tag{17-19}$$

or

$$Y = \sum_n [\text{MX}_n](a_{\text{MX}_n} - a_{\text{M}}) \tag{17-20}$$

Consider Y as a function of f; Y is a maximum when $dY/df = 0$ or

$$\frac{d}{df}\sum_n [\text{MX}_n](a_{\text{MX}n} - a_{\text{M}}) = 0 \tag{17-21}$$

This derivative is a complicated quantity, but in a number of limiting cases it can be considerably simplified. For example, if the formation constant for the first species is large compared to those of the second, third, and higher species, then the only species present besides M when MX_a is being formed will be MX_b. Now Eq. (17-20) becomes

$$Y = [MX_a](a_{MX_a} - a_M) + [MX_b](a_{MX_b} - a_M) \qquad (17\text{-}22)$$

and if we choose to measure the absorbance of this solution at a wavelength where $a_{MX_b} = a_M$, then Eq. (17-21) reduces to

$$Y = [MX_a](a_{MX_a} - a_M) \qquad (17\text{-}23)$$

and if a plot of Y versus f is made, the only contributing term will be $[MX_a](a_{MX_a} - a_M)$. Alternatively, if we choose a wavelength where $a_{MX_a} = a_{MX_b}$ then Eq. (17-21) becomes

$$Y = [MX_a](a_{MX_a} - a_M) + [MX_b](a_{MX_a} - a_M) \qquad (17\text{-}24)$$

or

$$Y = \{[MX_a] + [MX_b]\}(a_{MX_a} - a_M) \qquad (17\text{-}25)$$

and if it is assumed that MX_b is formed in an equilibrium reaction with MX_a, then the sum $[MX_a] + [MX_b]$ is a constant and any such reaction forming MX_b will have no effect on the value of Y. In either of these two cases, with $a_{MX_b} = a_M$ or with $a_{MX_a} = a_{MX_b}$, Y may be plotted versus f to yield a plot showing a maximum giving the value of $n = f/(1 - f)$.

Similarly, during the formation of the third species, there will be no appreciable concentration of free metal ion and the only species of importance will be MX_a, MX_b, and MX_c. If the formation constant of MX_a is large, its concentration will be given by

$$(1 - f)C - \sum_{n>a} [MX_n] \qquad (17\text{-}26)$$

which is completely analogous to the expression for the concentration of free metal ion [Eq. (17-4)]. Likewise, Y' is defined as

$$Y' = A_{obs} - (1 - f)a_{MX_a}C \qquad (17\text{-}27)$$

and a plot of Y' versus f at a wavelength where $a_{MX_c} = a_{MX_a}$ or at which $a_{MX_c} = a_{MX_b}$ will yield a curve at whose maximum the value of n for MX_b will be given by $f/(1 - f)$.

In a similar manner, all subsequent equilibria can be studied. In other words, to detect the maximum of a species MX_n, plot $A_{obs} - (1 - f)Ca_{MX_{n-1}}$ versus f at a wavelength where either $a_{MX_{n+1}} = a_{MX_{n-1}}$ or $a_{MX_{n+1}} = a_{MX_n}$.

Clearly, for Job's method to give meaningful data, certain conditions must hold: (1) Complexes of fairly high stability must be formed, otherwise the absorbance will be time-dependent and ill defined. (2) The stepwise stability constants must be large for the intermediate species to have any distinct existence; the difference between the K_n's must also be large for the assumption $MX_a = (1 - f)C$ to be valid. (3) The spectra of the several species should be as different as possible. (4) Isosbestic points must exist. (Isosbestic points are wavelengths where pairs of species have identical molar absorptivities.) Such points are determined experimentally by observing the wavelengths where the spectra of the two species in question are coincident. If these two spectra were recorded from solutions of the same concentration and path length, then the points where the spectra cross are isosbestic points.

DATA TO BE ACQUIRED AND CONTROL IMPLIED

From the discussion of the theory behind the method of continuous variations, the experimental procedure is obvious. One prepares a series of solutions in which the ratio of metal to ligand concentration is varied while the sum of these two concentrations is kept constant. Each of these solutions is then placed in the spectrophotometer and absorbances measured at the appropriate wavelengths. The data thus collected are then reduced and plotted as a function of the mole fraction of ligand, the maximum determined, and n calculated from the value of f at the maximum.

In order to get a meaningful variation in f, a fairly large number of different solutions (about ten) must be run and, depending on the number of distinct complex ions formed, absorbance data from each solution must be taken at a fair number of wavelengths. This procedure is ideally suited for execution under computer control, and a number of advantages accrue. Since each solution must be treated in the same way, i.e., data taken at the same wavelengths for each solution, the computer can be programmed to perform the same operations on each solution until all have been run. In a single-beam spectrophotometer, such as a Spectronic 20, each change of wavelength would require a readjustment of the blank or 0 absorbance reading. Under computer control a blank need only be run once. The computer can be interfaced and programmed to control the wavelength as well, so each solution need be handled only once. This prevents mix-ups due to repeated handling of many cuvettes. Finally, the machine can acquire the data much more rapidly than the student can, leaving the student more time to spend on data-reduction tasks and allowing him more time to ponder the chemistry involved.

DATA REDUCTION

Once acquired, the data must be reduced and plotted in order to determine the composition of the various nickel-ethylenediamine complexes.

The computer acquires only intensity values, and these must be converted to absorbance values before Y can be calculated. Absorbance is defined as

$$A = \log(I_0/I_t) \qquad (17\text{-}28)$$

where I_0 is the intensity of incident light and I_t is the intensity of transmitted light. The quantity I_0 is taken to be proportional to the blank value, and I_t is the value observed for the sample.

In the case of nickel–ethylenediamine complexes, three main species, MX_a, MX_b, and MX_c, are known to form. Isosbestic points exist at 662 nm for MX_a, 605 and 578 nm for MX_b, and 545 and 530 nm for MX_c.

It has been shown above that a plot of Y versus f will give a maximum at the point where the value of n can be calculated directly. Thus, the first step in analyzing the data is to plot the function $[A_{obs} - (1 - f)A_M]$ vs f for the data collected at 622 nm, where f is the mole fraction of the enthylenediamine in the solute. This plot will give the value of n for the first complex formed (MX_a).

In order to determine the value of n for the second species (MX_b), a plot of $[A_{obs} - (1 - f)Ca_{MX_a}]$ vs f, where C is defined by Eq. (17-2), must be constructed. Thus, the value of a_{MX_a} must be determined. To calculate a_{MX_a}, recall that as long as the stability constants for the reactions given in Eq. (17-12) are large and well separated, it can be assumed that wherever $[MX_a]$ is a maximum, the only species present are MX_a and X; all of the metal M has been reacted to form MX_a. Under these conditions, $[MX_a] = (1 - f)C$. To determine a_{MXa}, the value of f at which Y is a maximum on the 662-nm plot is noted. This is the value of f where $[MX_a]$ is a maximum. From the absorbance data taken at 605 nm and 578 nm on the solution where $[MX_a]$ is a maximum, a_{MX_a} at those wavelengths may be determined using Beer's law. Having calculated a_{MXa} at these two wavelengths, a plot of $[A_{obs} - (1 - f)Ca_{MX_a}]$ vs f for the absorbance data collected at 605 and 578 nm allows determination of n from the value of f at the maxima. Similarly, from the absorbance data taken at 545 and 530 nm on the solution where $[MX_b]$ is a maximum, a_{MX_b} can be calculated, and an appropriate plot will yield the value of n.

HARDWARE INTERFACE

In any on-line experiment, machine-to-machine communication is provided through an interface. In many cases, this communication need only be in one direction, usually from experiment to computer. However, the general-purpose computer is equipped to handle two-way communication as easily as one-way communication, and the use of the computer not only for data acquisition but also for experiment control is not only feasible but desirable. In the present experiment, it is desirable for the computer to be capable of controlling spectrophotometer-wavelength selection as well as handling the data acquisition.

The Spectronic 20 spectrophotometer is recommended for the present application on the basis of its low cost and simple, rugged design. Since, for many purposes, only a moderate scan speed is desired and high resolution is not necessary, a synchronous motor is employed to control the dispersive element in the colorimeter. To achieve a more drift-free signal from the phototube, the standard amplifiers in the instrument are bypassed completely, and a solid-state operational amplifier is substituted to provide the data signal to the computer.

The computer instrumentation consists of a laboratory computer system with interface modules such as that described at the start of Sec. II. Since the computer interface provides electronic communication between machines, another interface must be constructed to convert electrical signals generated by the computer to mechanical actions which control the monochromator. A block diagram of this interface is shown in Fig. 17-1.

It is desirable, if possible, to have the computer control both the direction and magnitude of rotation of the motor used. Since a synchronous motor is used, the amount of rotation is determined by the time the motor is allowed

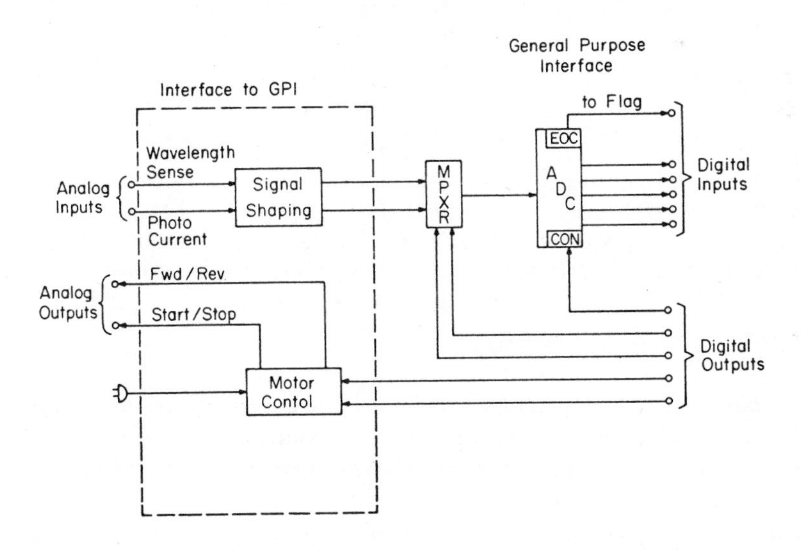

Fig. 17-1

Block diagram of the Spectronic 20 interface. [From *J. Chem. Educ.*, **50**, 428 (1973).]

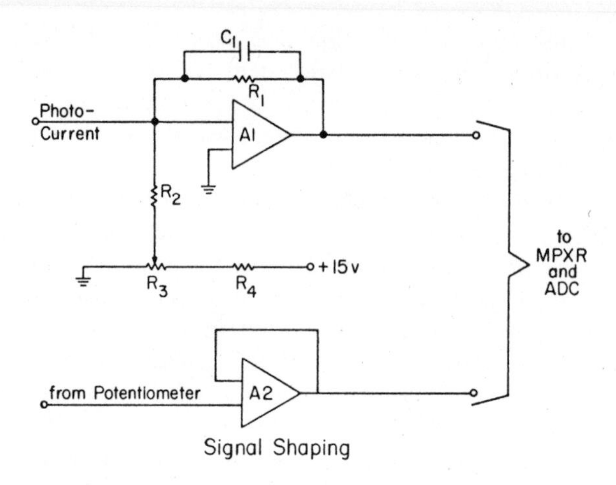

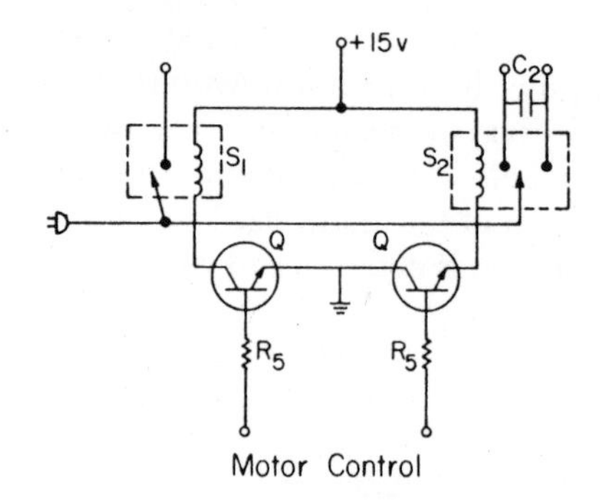

Fig. 17-2

Motor control and signal-shaping circuit for the Spectronic 20 interface. Parts list: S_1, Reed Relay (Magnacraft W105MPCX); S_2, Relay (Potter and Brumfield KA5DY); C_1, 5000 pF; C_2, 0.5 μF; R_1, 69.9 MΩ; R_2, 1 MΩ; R_3, 1 kΩ; R_4, 100 kΩ; R_5, 2 kΩ; Q, Transistor (Fairchild 2N3965); $A1$, $A2$ = Operational Amplifiers (SN72741). [From *J. Chem. Educ.*, **50**, 428 (1973).]

to rotate. Starting and stopping the motor are accomplished through activation of a clutch on the armature. Only two control signals are required and are used to actuate transistor relay drivers. These relays in turn start and stop the motor (engage or disengage the clutch) and determine the direction of rotation.

If the computer is to control the rotation of the grating in the Spectronic 20 in a useful manner, it must also be able to sense the position of the grating.

A potentiometer is therefore connected to the shaft of the monochromator through gears mounted on the monochromator shaft and on the potentiometer wiper. As the monochromator shaft is rotated, the potentiometer changes its resistance and acts as a voltage divider. Because the angle of rotation of the grating is linear with wavelength, the fraction of the applied voltage appearing at the armature of the potentiometer is related to the wavelength of the light passing through the cell. This analog signal must be digitized through the analog-to-digital converter (ADC). Because phototube data must also be acquired through the same ADC, these two signals (wavelength sense and phototube output) must be multiplexed. In order to control the multiplexer, two more control signals must be available to select the appropriate multiplexer channel. Thus, for control of the experiment, a total of four digital signals or bits are required: two for motor control and two for data-acquisition control.

As mentioned before, Fig. 17-1 shows a block diagram of the interface. The block called *motor control* consists of the two relay drivers and relays shown in Fig. 17-2. The motor which they control is a 4-rpm synchronous motor (Model AR-DA, Hurst Co., Princeton, Indiana, or similar). The multiplexer and ADC are part of the general-purpose laboratory computer system.

The signal-shaping part of the interface is shown also in Fig. 17-2. The signals to the multiplexer are the two control signals discussed previously. The analog input from the potentiometer is a varying voltage and is followed by operational amplifier *A*2, used in a voltage-follower mode. The other signal, the photocurrent, is taken directly from the phototube. Another operational amplifier, *A*1, is used to convert the photocurrent to a voltage in a range convenient for the ADC. The capacitor C_1 in Fig. 17-2 is used to limit the bandpass of this amplifier. The resistor network consisting of resistors R_2, R_3, R_4 is used to compensate for any dark current from the phototube. R_1, the feedback resistor, was chosen so that a photocurrent of approximately 150 nA will produce a -10-V output from the current-to-voltage converter.

SOFTWARE

The software required to operate this interfaced Spectronic 20 should supervise data acquisition as well as control the wavelength. For data acquisition, the computer must decide which datum (wavelength or transmittance) to take at any given moment in real-time, select the proper multiplexer channel, and acquire the datum.

If the datum is from the phototube, then no real-time calculations need be carried out, although some may be done. For example, if one wishes to

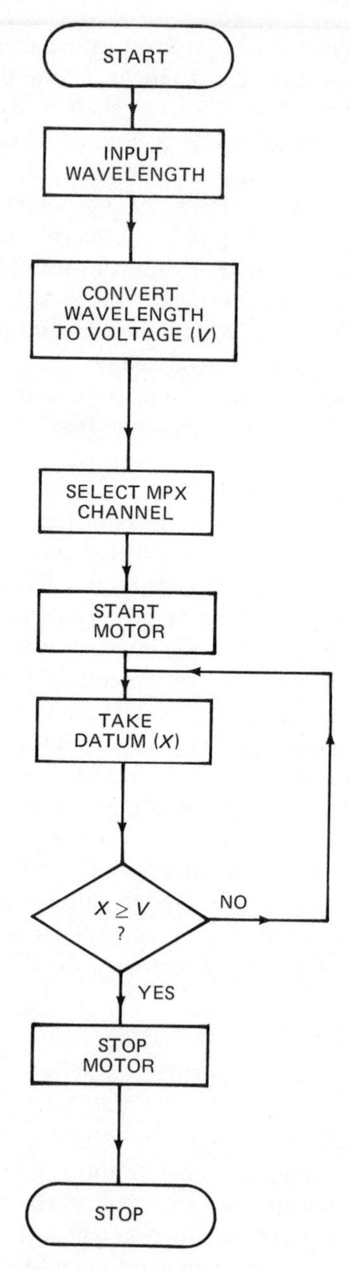

Fig. 17-3
Flow chart for a program to set monochromator to any wavelength requested. [From *J. Chem. Educ.*, **50**, 428 (1973).]

subtract out the blank values in real-time, this could be done as long as the data-acquisition rate is slow enough. Data collected from the phototube need to be corrected for blank values due to the nonlinear response of the phototube.

If the datum is from the monochromator, the computer must decide in real-time if the proper wavelength has been reached and output the proper control bits to stop the motor. In this case, the data-acquisition rate must be slow enough to permit whatever real-time processing is necessary to determine the position of the grating. Since the motor is relatively slow (4 rpm), the entire spectrum (300 nm) is scanned in approximately 6 sec. If a resolution of one part in 300 is required, then the ADC must sample at least 300 data points in those 6 sec (50 points per second). With these specifications in mind, a flow chart (Figure 17-3) for a program to set the monochromator to any wavelength desired can be drawn. Note that some real-time processing is necessary. At a 50-Hz data-acquisition rate, this processing can take no longer than 20 msec (which may allow execution of several BASIC statements).

In order to demonstrate the usefulness of this system and to illustrate the principles involved in the software necessary to operate the Spectronic 20 under computer control, consider the flow diagram of a program which would allow the instrument to be operated as a pseudo-double-beam recording spectrophotometer. This flow diagram is given in Fig. 17-4. The instrument becomes a pseudo-double-beam recording spectrometer in this manner; first a blank solution is scanned, and data are taken from the phototube at some fixed rate. The digitized values are saved in an array $B(I)$. Next, the sample solution is placed in the colorimeter and the spectrum scanned at the same data-acquisition rate as the blank. These data are saved as another array $X(I)$. After data acquisition is complete, the data are converted to percent transmittance and can be listed or plotted on the Teletype or the storage oscilloscope.

After an initial dialog between experimenter and computer wherein the wavelength monitor is calibrated and the number of data to be taken is determined, the machine asks that the blank be run and then waits for the console switch or other indicator to be set before scanning the blank. Next, data-acquisition hardware and software is initialized and a data-acquisition frequency (F) selected. Control bits are then used to select the proper multiplexer channel and to start the motor. Data are then collected from the ADC at intervals.

After data acquisition is complete, the program displays the data on the Teletype either as a plot of percent Transmittance vs wavelength or as a table of $\%T$ values. The reader is referred to the introduction to Section II for more details of the types of REAL-TIME BASIC control functions required to implement this flow chart.

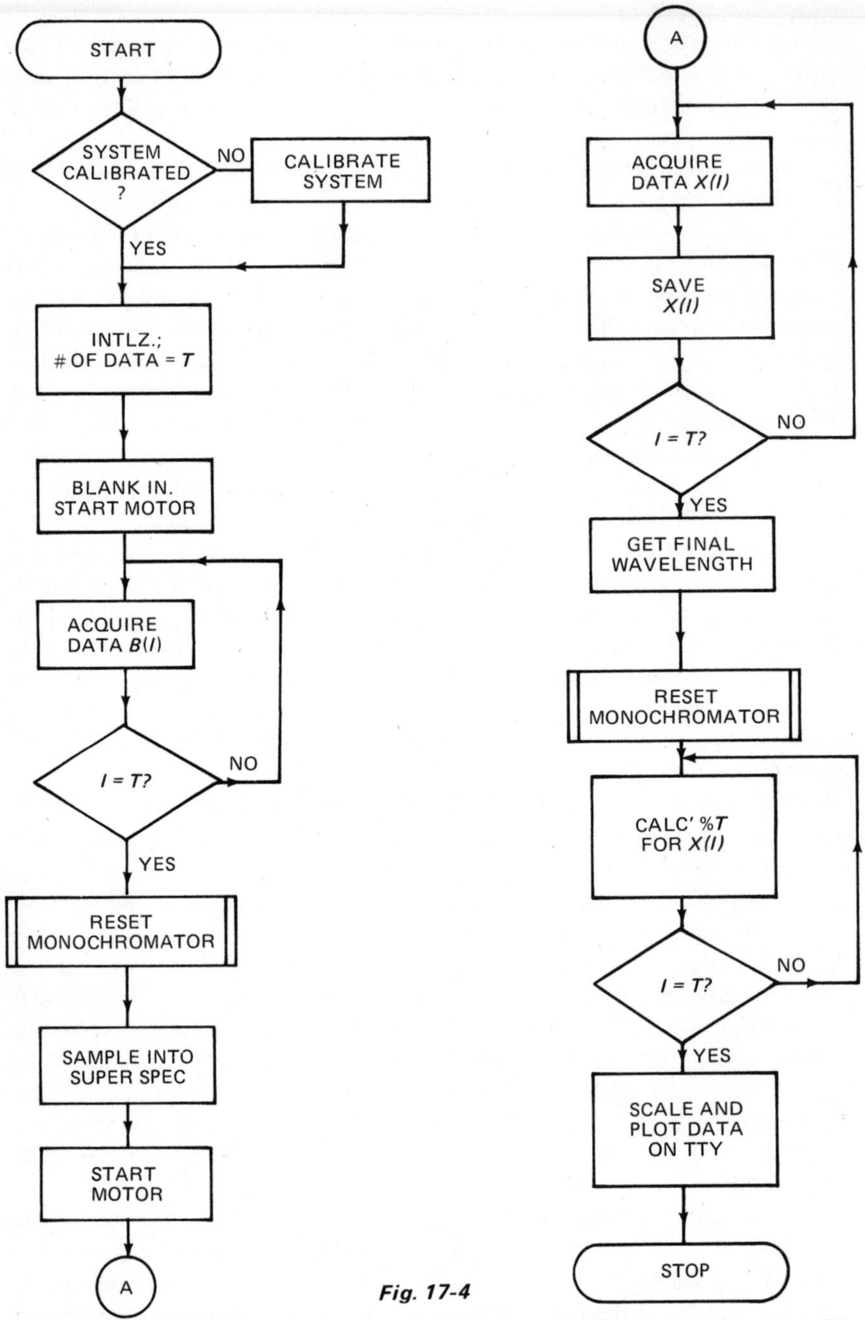

Fig. 17-4

Flow chart for a program to allow pseudo-double-beam operation of the Spectronic 20. [From *J. Chem. Educ.*, **50**, 428 (1973).]

With this system, a number of experiments in the chemistry laboratory can be envisioned. (1) Through computer control a simple colorimeter can be used as a pseudo-double-beam spectrophotometer, as above. (2) Since data-acquisition rates can be quite high if no real-time processing is involved (to tens of kilohertz, depending on the ADC used), time-dependent analysis of absorption peaks can be done for kinetics studies. (3) On a slightly slower time scale, time-dependent absorbance studies at multiple wavelengths can be done utilizing computer control of the monochromator. (4) Multiple data bracketing peak regions can be acquired, thus ensuring that analytical measurements are made at peak maxima. (5) For very weakly absorbing systems, signal averaging can be performed.

The experiment you will perform involves the use of Job's method (refs. E-14 and E-15) (the method of continuous variations) to elucidate the composition of some complexes formed between Ni(II) and ethylenediamine. The computer-controlled colorimeter is used to minimize the number of manual adjustments usually required, relieve the tedium of the experiment, and provide more effective data collection and handling. In addition, the computer can be programmed to process and display the data.

EXPERIMENTAL

Equipment Needed

1 Laboratory computer system with programmable clock and digital and analog I/O interfaces
1 Spectronic 20 modified as described above (to be previously assembled by instructor)
1 ± 15-V power supply
2 10-ml Mohr pipets
1 Optical cell for Spectronic 20
Assorted beakers

Chemicals

0.10 *M* Nickel sulfate
0.10 *M* Ethylenediamine

Procedure

A. Wavelength Setting and Calibration

1. Write a program to control calibration of the potentiometer on the wavelength dial of the Spectronic 20 and also to allow computer

selection of any specific wavelength desired. See the flow chart of Fig. 17-3. *Note*: These programs should be *written in segments* such that they can be combined into the large program needed to do the entire experiment. Therefore, avoid using the same labels and symbols for different functions in different program segments. It is also important to note that control of the spectrophotometer is generally accomplished through a call to an appended assembly-language subroutine. Thus, attention must be paid to the hardware configuration to determine exactly how the functions for motor control (start/stop, forward/reverse) and multiplexer channel selection (1 or 2) may be invoked.

2. Connect the Spectronic 20 modified as indicated above following the diagrams of Figs. 17-1 and 17-2. Note that the interface is *to be supplied by the instructor*; students are expected to connect the interface to the computer and spectrometer, but are not expected to assemble the interface.
3. Test the program(s) written in step 1.

B. Complete Job's Plot Program

1. Write a program to implement the flow charts in Fig. 17-4. Include data reduction and output of data on scope or Teletype. The program should also include a method of determining the maxima of the plots and calculating the corresponding value of n. (*Note*: Computer memory restrictions may require output of the raw data, with data reduction taking place in a second program.)
2. Prepare a series of solutions containing:
   (a) x ml of 0.1 M ethylenediamine
   plus
   (b) $10 - x$ ml of 0.1 M nickel sulfate
   where x varies from 3 to 9 ml in $\frac{1}{2}$ ml intervals. Prepare one solution containing only 10 ml of 0.1 M nickel sulfate and one solution containing only 10 ml of 0.1 M ethylenediamine.
3. Run your program. Recall that data must be collected at isosbestic points. For the nickel sulfate–ethylenediamine system, these wavelengths can be easily determined experimentally. For the first species (MX_a), make absorbance measurements at 622 nm; for the second species (MX_b), use 605 and 578 nm; for the third species (MX_c), measure at 545 and 530 nm.

Section III

Appendices

Appendix A

REAL-TIME BASIC (RTB) for Varian 620 Computers (Oregon/Nebraska Version)

INTRODUCTION

The BASIC programming approach described in this section may be implemented on Varian 620/i, 620/L, and 620/f computers. In addition, the Varian 73 computer can be used in the same way. Our development of REAL-TIME BASIC (RTB) for these systems has relied on simple modifications of the BASIC compiler as supplied by Varian. This compiler is a version of the language, originally developed at Dartmouth College, which includes all of the standard arithmetic operations as well as trigonometric functions and matrix-algebra routines. The standard Varian version of the language requires approximately seven thousand words of memory for the compiler, leaving the remainder available for user source code. A user may expand the available source-code area (thus allowing larger programs) by deletion of the matrix algebra and/or trigonometric function packages. This yields about 700 additional words of memory. Since this is a true interpretive compiler, all user source code and the compiler must be stored in core at run time. It is possible to modify the compiler to use a mass storage device (i.e., magnetic disk) to store user source code. At least one user has done this; however, such a procedure slows execution times and is not generally suitable for real-time experimentation. As released by Varian, BASIC contains no subroutines to support devices of the type needed in the laboratory (i.e., analog and digital I/O devices, control and sense lines, etc.). Fortunately, Varian BASIC does provide a CALL statement, which can be used to allow linkage of assembly language subroutines to a BASIC program. It is this feature that we have utilized to develop the RTB systems which are presently in use at the Universities of Oregon and Nebraska. In passing, we should note that Varian does offer a special hardware/software system, called ADAPTS, which includes an extended BASIC compiler with many of the I/O capabilities required for laboratory computing. This compiler is very similar to our version, but appears only to be available as part of an ADAPTS system. As you might infer from this brief introduction, the addition of assembly-language subroutines (necessary for use of RTB) will place further demands on the available memory and leave less space for storage of edited user source code. Although almost all experiments described in this manual can be performed using an 8-K word computer, we strongly recommend at least a 12-K word system for maximum flexibility. The reasons for this should become clear in the following paragraphs.

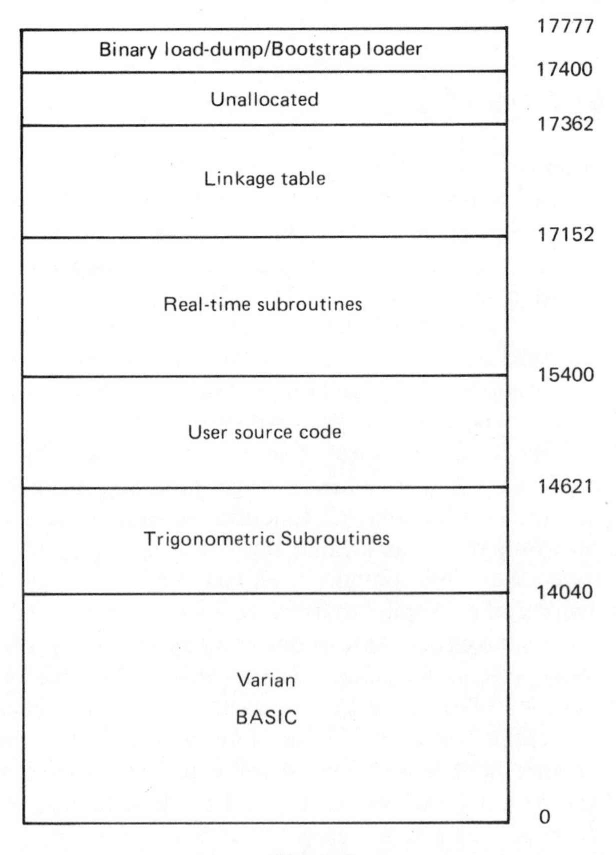

Fig. A-1
Memory map for an 8-K computer version of RTB with matrix algebra routines deleted.

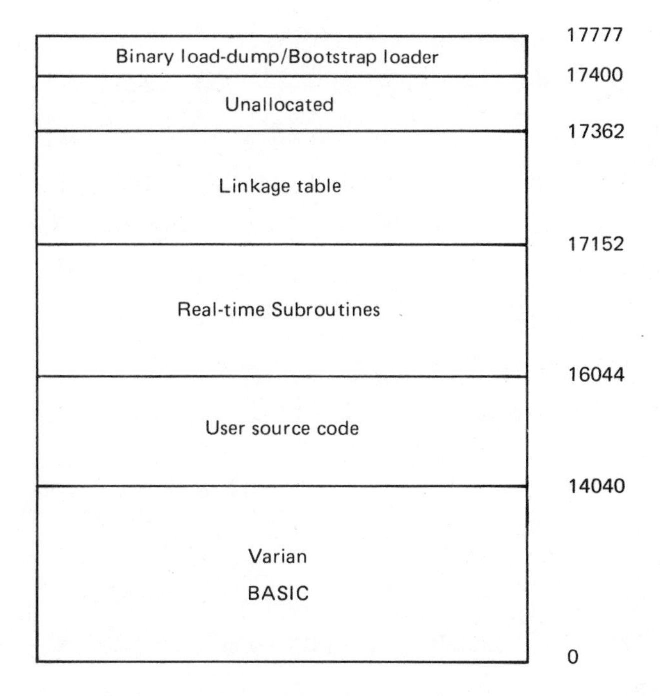

Fig. A-2
Memory map for an 8-K computer version of RTB with trigonometric and some oscilloscope display routines deleted.

MEMORY REQUIREMENTS AND SUBROUTINE LINKAGE

In order to add subroutines to the BASIC compiler, it is necessary to understand a few details of the compiler's construction. Linkage to the Varian BASIC interpreter is through a list containing five words per subroutine. This list is essentially a set of directory entries which stipulate name, entry point, and number of arguments for each subroutine. The compiler determines the location of this linkage table by examining the contents of memory location 10_8, which contains its address. Other important memory locations are the following: 11_8, which contains the lowest address used for user source-program storage, and location 22_8, which is reserved for the highest location used by the BASIC compiler. Obviously, the linkage table may conveniently be placed at either the top or bottom of available core memory. The relationships for one typical RTB configuration are depicted in the memory map of Fig. A-1. (The BASIC compiler shown here has been slightly *modified from the*

```
DEMO    ******  REAL-TIME BASIC SUBROUTINE PACKAGE  ******                           PAGE   1

14040                       1        ,ORG    ,014040    ORG LOCATION                                    00000003
                            2 * PEN(L)                  IF L IS POS. DROP THE PEN, IF NEG. LIFT IT      00000004
14040  006027 114060        3 PEN1   ,LDBE*  ,PEN       LOAD THE ADDRESS OF THE ARGUMENT                00000005
14042  044015               4        ,INR    ,PEN       BUMP THE RETURN POINTER                         00000006
14043  016000               5        ,LDA    ,0,2       SIGN BIT + EXPONENT + M.S. MANTISSA             00000007
14044  001002 014050        6        ,JAP    ,*+4       INSPECT THE SIGN BIT                            00000008
14046  100134               7        ,EXC    ,0134      IF L IS NEGATIVE, LIFT THE PEN                  00000009
14047  001006               8        ,DATA   ,01006     SKIP THE NEXT INSTRUCTION                       00000010
14050  100034               9        ,EXC    ,034       IF L IS POSITIVE, DROP THE PEN                  00000011
14051  006010 077777      10        ,LDAI   ,077777    FULL SCALE DELAY FOR PEN RESPONSE               00000012
14053  004260              11        ,LRLA   ,16        ROTATE                                          00000013
14054  005311              12        ,DAR    ,          DECREMENT                                       00000014
14055  001002 014053      13        ,JAP    ,*-2       WAIT UNTIL A REG GOES NEGATIVE                  00000015
14057  001000 000000      14        ,RETU   ,0         RETURN                                          00000016
       014060              15 PEN    ,EQU    ,*-1                                                       00000017
14061  001000 014040      16        ,JMP    ,PEN1      ALLOW FOR FORWARD RELATIVE ADDRESSING           00000018
14063  120240              17 SU01   ,DATA   ,'   PEN'.1,PEN   LINKAGE TABLE ENTRY NO. 1                00000019
14064  120320
14065  142716
14066  000001
14067  014060
14070  014076              18        ,DATA   ,SU02      <------LINK 1                                   00000020
                           19 * START                   START THE CLOCK + CLEAR AND START THE ADC.      00000021
14071  000000              20 STRT   ,ENTR   ,                                                          00000022
14072  100227              21        ,EXC    ,0227      RESET THE CLOCK                                 00000023
14073  100031              22        ,EXC    ,031       CLEAR AND START THE DSI ADC                     00000024
14074  001000 114071      23        ,RETU*  ,STRT      RETURN                                          00000025
14076  120323              24 SU02   ,DATA   ,' START',0,STRT  LINKAGE TABLE ENTRY NO. 2                00000026
14077  152301
14100  151324
14101  000000
14102  014071
14103  014121              25        ,DATA   ,SU03      <------LINK 2                                   00000027
                           26 * WAIT(S)                 READ THE DIO SWITCHES AND RETURN THE VARIABLE   00000028
                           27 *                         "S".  WAIT IF THE MOST SIGNIF BIT IS SET.       00000029
14104  024011              28 WAI1   ,LDB    ,WAIT      LOAD THE ARGUMENT ADDRESS POINTER               00000030
14105  044010              29        ,INR    ,WAIT      BUMP THE RETURN POINTER                         00000031
14106  100144              30        ,EXC    ,0144      ENABLE THE DIO SWITCHES                         00000032
14107  102544              31        ,CIA    ,044       READ THE DIO SWITCHES                           00000033
14110  005000              32        ,NOP    ,                                                          00000034
14111  001004 014107      33        ,JAN    ,*-2       CONTINUE READING IF MOST SIGNIF. BIT IS SET     00000035
14113  002000 014173      34        ,CALL   ,HIN       FLOAT AND ASSIGN THE READING TO THE VARIABLE S  00000036
14115  001000 000000      35        ,RETU   ,0         RETURN                                          00000037
       014116              36 WAIT   ,EQU    ,*-1                                                       00000038
14117  001000 014104      37        ,JMP    ,WAI1      ALLOW FOR FORWARD RELATIVE ADDRESSING           00000039
14121  120240              38 SU03   ,DATA   ,'  WAIT',1,WAIT  LINKAGE TABLE ENTRY NO. 3                00000040
14122  153701
14123  144724
14124  000001
14125  014116
```

Fig. A-3

Sample RTB subroutine package.

original Varian version, and requires slightly more core.)

In the version shown in Fig. A-1, location 10_8 would contain the address 17152_8; location 11_8, the address 14621_8; and location 22_8, the address 15377_8. As is readily seen, there is a good deal of restriction on the size of a user's program when the real-time subroutines of this version are present (i.e., 367_{10} words of memory available).

However, additional memory may be freed by a number of simple steps. Obviously, if certain subroutines are not required by a particular program, it is foolish to reserve space for them if the space could be used otherwise (i.e., for program statements). For example, if the trigonometric subroutines were not required, this space could be released by changing memory word 11_8 to 14040 (the first address in the trig package). Furthermore, in the particular set of real-time subroutines used, certain oscilloscope display subroutines occupy the lower parts of the package. By changing memory

```
DEMO    ******  REAL-TIME BASIC SUBROUTINE PACKAGE  ******                    PAGE   2

14126  014153              40          ,DATA    ,SU04      <------LINK3                                      00000042
                           41 * DAC(C,D)                   OUTPUT "D" MILLIVOLTS FROM DAC CHANNEL "C".       00000043
14127  024020              42 DAC1     ,LDB     ,DAC       LOAD THE POINTER TO THE ARGUMENT "C"              00000044
14130  044017              43          ,INR     ,DAC       BUMP THE RETURN POINTER                           00000045
14131  002000 014161       44          ,CALL    ,HER       FIX "C"                                           00000046
14133  006150 000007       45          ,ANAI    ,7         LIMIT THE POSSIBILE NO. OF CHANNELS               00000047
14135  004246              46          ,LRLA    ,6                                                           00000048
14136  006110 100030       47          ,ORAI    ,0100030   FORM THE SELECT INSTRUCTION                       00000049
14140  054005              48          ,STA     ,DAC2      SAVE IT                                           00000050
14141  024006              49          ,LDB     ,DAC       LOAD THE POINTER TO THE ARGUMENT "D'              00000051
14142  044005              50          ,INR     ,DAC       BUMP THE RETURN POINTER                           00000052
14143  002000 014161       51          ,CALL    ,HER       FIX "D"                                           00000053
14145  103130              52          ,OAR     ,030       OUTPUT "D" TO THE DAC BUFFER                      00000054
14146  100030              53 DAC2     ,EXC     ,030       SELECT THE CHANNEL "C"                            00000055
14147  001000 000000       54          ,RETU    ,0         RETURN                                            00000056
       014150              55 DAC      ,EQU     ,*-1                                                         00000057
14151  001000 014127       56          ,JMP     ,DAC1      ALLOW FOR FORWARD RELATIVE ADDRESSING             00000058
14153  120240              57 SU04     ,DATA    ,'   DAC',1,DAC  LINKAGE TABLE ENTRY NO. 4                   00000059
14154  120304
14155  140703
14156  000001
14157  014150
14160  000000              58          ,PZE     ,          <------LINKAGE TABLE END FLAG                     00000060
                           59 *                                                                              00000061
       013622              60 $QS      ,SET     ,013622    =ENTRY ADDR FOR FLOAT                             00000062
       013702              61 $HS      ,SET     ,013702    =ENTRY ADDR FOR FIX                               00000063
                           62 *                                                                              00000064
                           63 * HER                        CONVERT THE FLOATING POINT VARIABLE POINTED TO    00000065
                           64 *                            BY THE B REG TO AN INTEGER AND RETURN WITH IT     00000066
                           65 *                            IN THE A REG AND STORED AT THE SAVE LOCATION.     00000067
14161  000000              66 HER      ,ENTR    ,                                                            00000068
14162  026000              67          ,LDB     ,0,2       LOAD THE ADDRESS OF THE VARIABLE                  00000069
14163  016000              68          ,LDA     ,0,2       LOAD THE SIGN BIT + EXPONENT + MS MANTISSA        00000070
14164  026001              69          ,LDB     ,1,2       LOAD THE LS MANTISSA                              00000071
14165  002000 013702       70          ,CALL    ,$HS,SAVE  USE THE COMPILER ROUTINE $HS TO FIX IT            00000072
14167  014172
14170  001000 114161       71          ,RETU*   ,HER       RETURN WITH THE NO. IN THE A REG                  00000073
14172  000000              72 SAVE     ,PZE     ,          INTEGER ALSO STORED HERE                          00000074
                           73 *                                                                              00000075
                           74 * HIN                        FLOAT AND ASSIGN THE FIXED POINT NO. IN THE       00000076
                           75 *                            A REG TO THE VARIABLE POINTED TO BY THE B REG.    00000077
14173  000000              76 HIN      ,ENTR    ,                                                            00000078
14174  026000              77          ,LDB     ,0,2       LOAD THE ADDRESS OF THE VARIABLE                  00000079
14175  064002              78          ,STB     ,*+3       STORE IT AS AN ARG FOR $QS                        00000080
14176  002000 013622       79          ,CALL    ,$QS,0     FLOAT THE NO. IN THE A REG                        00000081
14200  000000
14201  001000 114173       80          ,RETU*   ,HIN       RETURN                                            00000082
       014203              81 FWAV     ,EQU     ,*         FIRST WORD AVAILABLE FOR USER SOURCE CODE         00000083

                           83 * ADJUST THE INTERPRETER POINTERS                                              00000085
00010                      84          ,ORG     ,010                                                         00000086
00010  014063              85          ,DATA    ,SU01      ADDR OF THE FIRST MEMBER OF THE LINKAGE TABLE     00000087
00011  014203              86          ,DATA    ,FWAV      FIRST WORD AVAILABLE FOR USER SOURCE CODE         00000088
       000002              87          ,END     ,2                                                           00000089
```

```
DEMO   ******  REAL-TIME BASIC SUBROUTINE PACKAGE  ******

  0 INDIRECT POINTERS

  0 LITERALS

 18 SYMBOLS, ALPHABETICALLY WITH CROSS REFERENCE

$HS      013702     014165
$QS      013622     014176
DAC      014150     014127  014130  014141  014142  014157
DAC1     014127     014151
DAC2     014146     014140
FWAV     014203     000011
HER      014161     014131  014143  014170
HIN      014173     014113  014201
PEN      014060     014040  014042  014067
PEN1     014040     014061
SAVE     014172     014167
STRT     014071     014074  014102
SU01     014063     000010
SU02     014076     014070
SU03     014121     014103
SU04     014153     014126
WAIT     014116     014104  014105  014125
WAI1     014104     014117
```

word 22_8 to 16043 (the last address in these routines), some of the oscilloscope-display capabilities could be sacrificed and additional space would be made available for user source code (now a total of 1028_{10} words). The revised memory map is shown in Fig. A-2.

Now, consider the details of the directory entries which are required for each subroutine. The first three words are reserved for the name of the subroutine. Since two ASCII characters may be stored in each word, this allows subroutine names of up to six characters. The fourth word contains the number of arguments the subroutine accepts or passes, and the fifth word contains the address of the subroutine's entry point. According to the Varian BASIC convention, setting the first word of the name field to zero signifies the end of the list. The six-character subroutine name is always right justified (i.e., with leading ASCII blanks in the appropriate places if less than six alphabetic characters are used). When the BASIC compiler encounters a CALL statement, it first examines the contents of location 10_8 to find the address of the linkage table and then begins a sequential search of the table for an exact match for the subroutine name specified. This search continues until either the name is found or a zero in the first word of a name field is encountered. If the name is not located, an error message will result. The number of arguments in the CALL statement is checked, if the name is found. If this number does not correspond to that specified in the linkage list, then an error message results. The two error messages associated with the linkage search procedure are:

Error 16: Called routine does not exist
Error 17: Wrong number of parameters in CALL statement

If no error is found, the compiler then goes to the specified subroutine.

In our early versions of REAL-TIME BASIC, we followed the method discussed above to add the subroutines. However, the disadvantage of this linkage procedure is that the linkage table must occupy contiguous locations. Recently, we have modified the linkage convention slightly (this requires modification of the standard Varian BASIC compiler) to allow a different arrangement of the linkage list. In the modified version, a zero in the first word of a name field indicates the end of the list (as before) and a negative number in the first word of the name field indicates a valid name (as before, since all ASCII characters cause bit 15 to be a one, a negative number, by Varian convention). A positive number, however, indicates the address of the next entry in the linkage table. Thus, entries no longer need be contiguous (although they must be in a protected portion of memory so as not to be overwritten by user source code). This allows far more flexibility in addition of assembly-language subroutines, although the standard Varian method is certainly quite workable. Notice that this modification requires six words of memory per entry if the directory is not placed together as a block.

A listing of a DAS assembly for four REAL-TIME BASIC subroutines is contained in Fig. A-3. Three of them, START, WAIT, and DAC, are standard types of routines discussed in Sec. II and tabulated in Table II-1. The other, PEN, causes raising and lowering of a recorder pen.

As we have described above, this assembly yields the required code to add four subroutines, PEN, START, WAIT, and DAC, to the BASIC compiler. The linkage-table entries are not contiguous, but are dispersed through the code using the technique previously described. The entries are found at the beginning at addresses 14063_8, 14076_8, 14121_8, and 14153_8, respectively. The subroutines and linkage table resulting from the assembly above occupy locations 14040_8 to 14202_8 inclusive, and the previously mentioned pointers in locations 10_8 and 11_8 are set to reflect this fact. The assembled version of the code listed above is loaded after first loading the Varian BASIC compiler and deleting the matrix and trigonometric packages. The RTB compiler depicted in the memory map of Fig. A-4 then results.

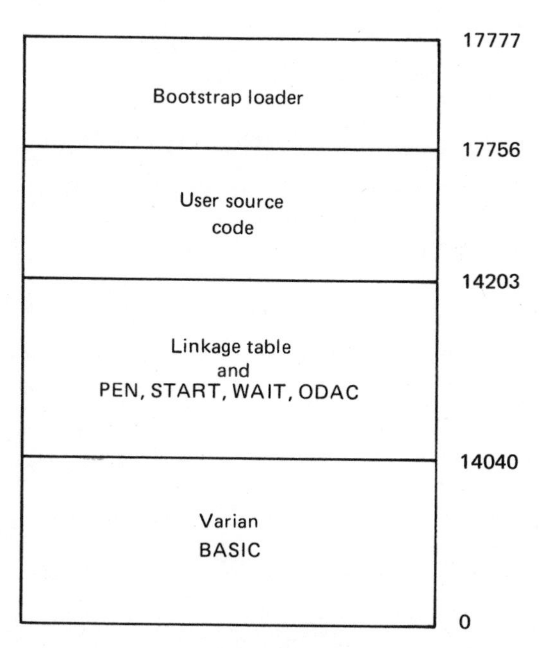

Fig. A-4
Memory map for an 8-K computer version of the sample RTB.

Appendix B

Purdue REAL-TIME BASIC

The high-level programming language for on-line experimentation developed at Purdue University was built around the BASIC language and will be referred to as Purdue REAL-TIME BASIC (PRTB). The software was developed by modifying the BASIC Interpreter available from Hewlett-Packard (H.P. 20392). The modifications generated have involved the development of a series of machine-language subroutines which are directly callable from the BASIC software and which are designed to communicate in a variety of ways with experimental systems. The characteristics of the PRTB data-acquisition and control software are exactly analogous to those described in Sec. II.

As normally provided by Hewlett-Packard, the BASIC software for a stand-alone (single-user) computer does not include instructions for controlling data acquisition or other devices for on-line experimentation. These subroutines must be written by the user and added to the H. P. BASIC software.

In this appendix we will describe some general considerations for modifying the BASIC software available from Hewlett-Packard in order to generate a modified BASIC for on-line experimentation.

The discussions presented here apply to a paper-tape-oriented computer system for single-user operation. The specific details will vary somewhat depending on the total configuration of the system, the current version of H. P. BASIC used, and the nature and length of subroutines to be added. For an accurate treatment of the reader's particular system, the manufacturer's specific documentation and recommendations should be followed.

The linkage format and the steps involved in the execution of an auxiliary machine-language subroutine using the Hewlett-Packard BASIC language are illustrated in Fig. B-1. Each subroutine is designated by a unique integer number from 1 to 63_{10}. The CALL statement must first specify the subroutine number; subsequently, several variable or constant parameters may be included in the CALL statement. For example, if subroutine 5 is a data-acquisition segment requiring certain initialization information, like the number of data points to take, the clock frequency, and the address of the

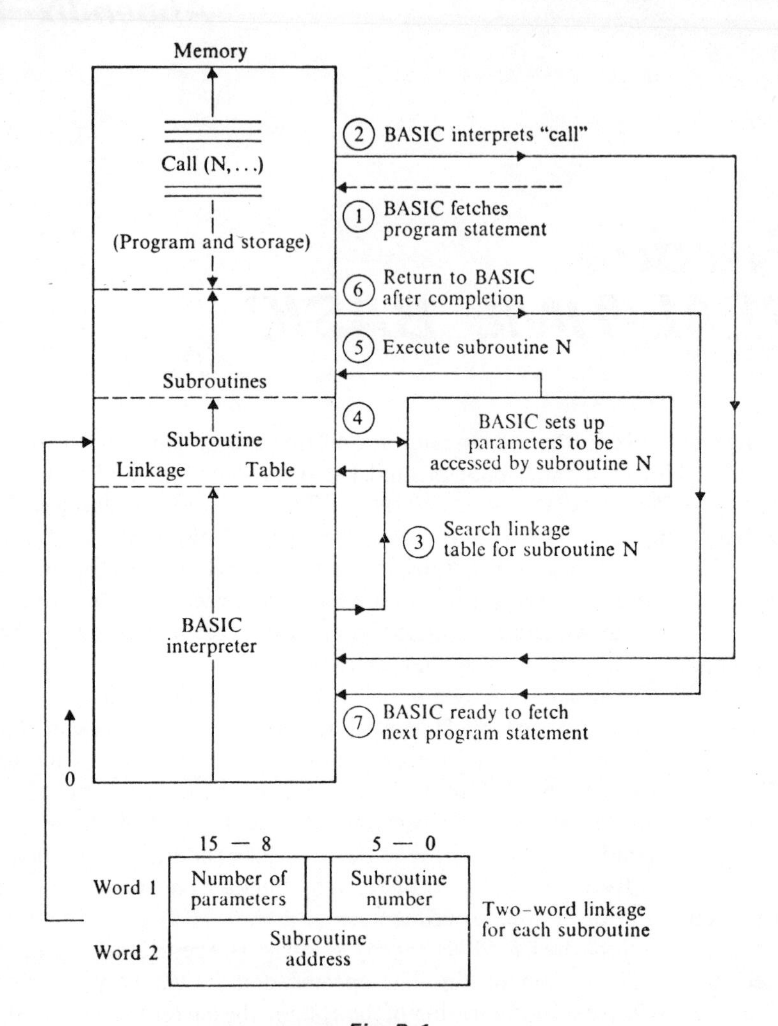

Fig. B-1

Linkage and execution of auxiliary subroutines with H. P. BASIC. [From Reference A-20, Chapter 11.]

first word of the data-storage buffer, these parameters could be transferred through the CALL statement as shown below:

$$\text{CALL}\ (5, T, F, A(1))$$

Before executing subroutine 5, BASIC will evaluate each parameter and set up an *address stack*. Upon entering the subroutine, the address of the last location in the stack will be in the A register. The stack itself will contain the addresses of each of the parameters in the CALL statement. Thus, the *A* register initially contains the address of an address. The last location in the stack contains the address of the first parameter, the next to the last contains the address of the second parameter, etc. Thus, the subroutine should be written to obtain parameters or addresses through the *A*-register linkage. The segment below illustrates what the initialization part of our data-acquisition subroutine might look like:

```
DATAC   NOP
        STA   B          SAVE STACK POINTR IN "B" REG.
        LDA   B, I       GET 1ST PARAMTR ADRSS
        STA   TEMP       SAVE ADRSS
        LDA   A, I       GET 1ST WORD OF FL. PT. VALUE
        STA   DCNT       SAVE 1ST PRT.
        LDA   TEMP       GET ADDRSS
        INA              INCRMNT TO GET 2ND WRD
        LDA   A, I       GET 2ND WRD OF FL. PT. VALUE
        STA   DCNT + 1   SAVE 2ND PRT
        ADB   DCRMT      DECRMNT STACK POINTR
        LDA   B, I       GET 2ND PARAMTR ADRSS
        STA   TEMP       SAVE
        LDA   A, I       GET 1ST WRD OF FL. PT. VALUE
        STA   CLOK       SAVE 1ST PRT
        LDA   TEMP       GET ADRSS
        INA              INCRMNT TO GET 2ND WRD
        LDA   A, I       GET 2ND WRD
        STA   CLOK + 1   SAVE 2ND PRT
        ADB   DCRMT      DECREMENT STACK POINTER
        LDA   B, I       GET 3RD PARAMTR ADRSS
        STA   PBFFR      SAVE AS POINTR FOR LATER DATA
              .          STORAGE
              .          .
              .          .
        JMP   DATAC, I
DCRMT   OCT   -1         DATA ACQUISITION AND STORAGE
DCNT    BSS   2          .
CLOK    BSS   2          .
PBFFR   OCT   Ø
A       EQU   Ø          SET A,B=0,1; ADRSS'S OF A,B REGS.
B       EQU   1
```

Note: When numerical data are transferred, the data must be handled as two-word values. This is because all BASIC variables are represented in floating-point form. In order to use the numerical values within the subroutine, they may have to be "integerized." Also, data transferred back to BASIC will have to be "floated." Conversion from floating-point to integer values (or vice versa) can be accomplished by calling standard subroutines. (Often the desired service subroutine is contained within BASIC and can be utilized. For example, the IFIX subroutine to convert floating-point to integer values can be called directly from the auxiliary subroutine once the programmer knows where IFIX is located in his particular version of BASIC.)

DETAILS FOR ADDING SUBROUTINES TO H. P. BASIC*

Important Memory Locations

The first word of available memory (FWAM) is contained in location 110_8 in the H. P. BASIC system. The last word of available memory (LWAM) is contained in location 111_8 in the H. P. BASIC system. The address of the first word available in the base page is contained in location 114_8. All locations from the location referenced in 114_8 to 1777_8 are not used by BASIC and are therefore available for CALL subroutines or other modifications.

For ease in user modification, locations 201_8 to 322_8 contain links to various subportions and subroutines of BASIC often used in creating customized systems. The identity and locations of these links is fixed (will not change with subsequent versions), but the contents of these locations are subject to change if the routines they point to move as a result of future revisions. The assembly-language listing of the H.P. BASIC interpreter captions each link briefly. Since these links are an integral part of BASIC, the user is responsible for interpreting and using this information.

Linkages to Subroutines

BASIC accesses through a table containing linkage information(called subroutines). Entries in the table, one per subroutine, are two words in length. Bits 5–0 of the first word contain the number identifying the subroutine (chosen freely from 1 to 77_8 inclusive), and bits 15–8 contain the number of parameters passed to the subroutine. (CALL statements with an incorrect number of parameters are rejected by the syntax analyzer.) The second word contains the absolute address of the entry point of the subroutine. (Control is transferred via a JSB.) Although subroutine numbers need not be assigned in any particular order, all entries in the table must be contiguous. An acceptable auxiliary tape contains the following:

1. An ORG statement to originate the program at an address greater than that of the last word of the BASIC system. The address of this last word +1 is contained in the location 110_8 of the standard

*Taken in part from *A Pocket Guide to the 2100 Computer* (Hewlett-Packard Co., Cupertino, California, September 1972 edition, by permission of copyright owner).

BASIC system. Hence, a suitable lower limit for the origin address can be determined by loading BASIC and examining location 110_8.
2. The subroutine linkage table described above.
3. The assembly language subroutines.
4. Code to set the following linkage addresses:
   a. In location 110_8 put the address of the last word +1 used in the auxiliary tape.
   b. In location 121_8 put the address of the first word of the subroutine linkage table.
   c. In location 122_8 put the address of the last word +1 of the subroutine linkage table.

Assuming, for example, that location 110_8 of the standard BASIC systems contains 13142_8, an acceptable auxiliary tape could be assembled from the following code:

```
         ORG   13142B
SBTBL    OCT   24Ø6       Subroutine 6 has 5 parameters
         DEF   SB6
         OCT   1421       Subroutine 17 has 3 parameters
         DEF   SB17
ENDTB    EQU   *
SB6      NOP
               :          Subroutine #6 body
         JMP   SB6,I
SB17     NOP
         :                Subroutine #17 body
         JMP   SB17,I
LSTWD    EQU   *
         OEG   11ØB
         DEF   LSTWD
         ORG   121B
         DEF   SBTBL
         DEF   ENDTB
         END
```

Acceptable calls to subroutines SB6 and SB17 might be

CALL (6, A, B, 1, N*3, SIN($X + Y$))
CALL (17, A(1), 5, N)

Note: Location 111_8 of the standard BASIC system contains the address of the last word of available memory. It is not possible to create a system which requires more space than that existing between the addresses in locations 110_8 and 111_8. Systems using all or most of this space leave very little space for the user of the system.

Deleting Matrix Subroutines from H. P. BASIC

This assembly-language pseudo-program shows a method of deleting the MAT execution package to gain more user space, or for replacing it with CALL routines or other customized code.

ORG ⟨contents of $21Ø_8$⟩
OCT Ø,Ø
ORG 11ØB
DEF ⟨contents of 211_8⟩

This sequence has the effect of preventing the syntax processor from recognizing "MAT" and of resetting the first word of available memory pointer to the first word of the matrix-execution package.

Core Map	Absolute addresses
B. B. Loader	17000_8
Available user program space	(12711_8) w/o Matrix 14314_8
User subroutines	(12032_8) w/o Matrix 13435_8
Subroutine linkage table	(12000_8) w/o Matrix 13403_8
H. P. BASIC	2000_8
Available user base page space	1734_8
H. P. BASIC	2_8

Fig. B-2

Typical allocation of 8-K core. Memory version of H. P. BASIC modified with PRTB subroutines, shown with and without overlaying matrix instructions.

TYPICAL CHARACTERISTICS OF PRTB SOFTWARE

As pointed out earlier, the specific features of a modified-version H. P. BASIC will depend on many variables, including core memory size, current version of standard H. P. BASIC software, nature of computer peripherals, the length and number of added subroutines, and whether or not BASIC matrix instructions are retained. A core memory map for a typical version of PRTB configured at Purdue is shown in Fig. B-2. Note the difference in user-program space available in the 8-K version with and without matrix instructions deleted.

This version of PRTB included 13 subroutines functionally identical to those described in Sec. II. The total amount of core memory required (including the linkage table) was 710_8 (456_{10}) locations. A more extensive subroutine package, which includes subroutines for graphics and alphanumeric displays on an interfaced storage oscilloscope, requires $1{,}553_8$ (875_{10}) memory locations. In this latter version, very little space is left for users' programs in 8 K of memory. Thus, it is recommended for a 12-K or larger memory machine.

Appendix C

General-Purpose Interface and Logic Experiment Devices

LOGIC DEVICES

A number of relatively inexpensive and versatile logic laboratory devices are commercially available in preassembled or "kit" forms. Most of these include, at a minimum, provisions for easy patchcord wiring of logic circuits, power supplies, a clock and two or more indicator lights and push buttons or switches. The price of such items is from $100 or less to well above $1,000. As one might expect, the more elaborate the system, the higher its cost. Because of the great variety of options available, a detailed list of specifications will not be included here. Rather, the *minimal* system will be described. This is adequate for performance of the digital logic experiments (Sec. I, Experiments 1–5 and Sec. II, Experiments 6 and 7), although the more elaborate (and more expensive) alternatives will certainly be equally suitable. One such device is the EL Instruments "Digi-designer." This unit, available in kit or assembled form, *does not include the integrated circuit chips necessary for the experiments* (which can be purchased separately), but does include the following:

Power supply: Fixed 5 V, 400 mA (with 5% regulation)
Clock: Six frequencies. Nominally, 1, 10, 100, 1000, 10,000, and 100,000 Hz
Pulsers: Two bounce-free push-button pulsers
Input/output connections: Four 5-way binding posts, one BNC connector, and 29 solderless 22 gage interconnecting pins
Logic lamps: 4
Logic switches: 4
Spring-type socket breadboard and case, price (late 1974): $99.95 assembled, $64.95 in kit form

More elaborate and exceedingly useful devices sold by the same manufacturer are the "Elite" series of breadboards:

Elite 1: $695 (late 1974)
Elite 2: $1,395 (late 1974)
Elite 3: $400 (late 1974)

Since both prices and specifications change rapidly, it is advisable to contact the manufacturer directly for current information on these matters. A great many other options are available from other manufacturers, such as the Digital Equipment Corporation and EL Instruments, Inc. Both of these offer extensive digital-logic laboratory supplies.

The addresses of these manufacturers are:

Digital Equipment Corporation
146 Main Street
Maynard, Massachusetts 01754
Attention: Logic Products Sales Support Manager

EL Instruments, Inc.
61 First Street
Derby, Connecticut 06418

GENERAL-PURPOSE COMPUTER INTERFACES

The introduction of Sec. II stipulates the functional requirements of a general-purpose interface system adequate to allow performance of the experiments in this manual. For special devices offered by a particular computer manufacturer for use with specific computer systems, the best source of information is the manufacturer. It is recommended that, prior to purchasing any general-purpose interface from another source, you check with the manufacturer of your computer in regard to his capability to provide such an item.

Real-time data acquisition and control systems are available from a number of manufacturers and greatly simplify the interfacing task for users of computers who can afford to take advantage of a general-purpose system. Fundamentally, the approach is to provide an interface of a general peripheral device to a variety of computers. This general device usually contains prewired electronic card slots, necessary power supplies, and a mounting enclosure. The user selects from a variety of options those devices suited to his needs. The appropriate cards (which are the same, regardless of the computer in question) are simply plugged into the interfaced general device. This makes it possible for a single manufacturer to offer a general-purpose system for a wide variety of computers. Because of the generality of the approach, it *may* be somewhat more expensive. However, because of its flexibility, it is highly recommended.

As with the logic systems discussed above, prices and specifications vary greatly, so the prospective user should contact the manufacturer for detailed information regarding the availability and cost of systems for his particular computer. Two such systems are:

CAMAC Crates and Accessories, available from:
Standard Engineering Corporation
44800 Industrial Drive
Fremont, California 94538

Real-Time Peripheral Devices, available from:
Computer Products
Post Office Box 23849
Ft. Lauderdale, Florida 33307

Data-acquisition systems are also available from Datel Systems, Incorporated. This manufacturer offers a wide variety of devices suitable for use in computer-assisted experimentation. Datel's address is:

Datel Systems, Inc.
1020 Turnpike Street
Canton, Massachusetts 02021

Appendix **D**

Equipment and Reagents List

LOGIC EXPERIMENTS

TTL series 7400 dual-in-line packages are usually specified. Instructors may optionally substitute alternate equivalent components.

Experiment 1

1 EL Digi-designer (see Appendix C)
1 5-V variable power supply
1 Quad dual-input AND gate package (SN 7408)
1 Quad dual-input NAND gate package (SN 7400)
1 Quad dual-input OR gate package (Signetics SP 384A)
1 Quad dual-input NOR gate package (SN 7402)
1 Sine-wave generator
1 Volt–ohm multimeter
1 Oscilloscope (dual-trace works best)
2 Variable resistance boxes (15 Ω–10 MΩ)

Experiment 2

1 EL Digi-designer
2 Triple 3-input NAND gates (SN 7410)
1 Quad 2-input NOR gate package (SN 7402)
1 Quad 2-input OR gate package (Signetics SP 384A)
6 Quad 2-input NAND gate packages (SN 7400)
1 Hex inverter package (SN 7404)
1 Pulse generator (0 to 5 V), variable frequency
1 Storage oscilloscope

Experiment 3

1 EL Digi-designer
2 Dual-in-line JK flip-flop packages (SN 7476)—(4 for option F4)
1 Quad 2-input NOR gate package (SN 7402)
2 Quad 2-input AND gate packages (SN 7408)
1 Hex inverter package (SN 7404)
1 Quad 2-input OR gate package (Signetics SP 384A)
2 NAND gates for each switch to be buffered (usually one 7400 package is sufficient)
1 Dual-trace regular oscilloscope OR
1 Dual-trace storage oscilloscope (should have scan rate of at least 100 nsec/div.)
1 Variable resistance substitution box (15 Ω → 10 MΩ)
1 Variable capacitance substitution box (0.0001 → 0.22 μ)

Experiment 4

1 EL Digi-designer
1 ±15-V power supply
4 Voltage reference sources (variable 0 to 15 V)
1 Field-effect-transistor switch (Teledyne Crystalonics CAG-30. If a substitute is used, make sure that the gate operates on normal TTL logic levels.)
1 Voltage comparator (LM 710C)
2 BCD counters (SN 7490)
1 Operational amplifier (SN 72741)
1 Function generator (see Appendix E)
1 Pulse generator (variable rate)
1 Dual-trace storage oscilloscope
1 Volt–ohm multimeter
1 Toggle switch
5 Capacitors: 5 pF, 50 pF, 500 pF, and two 0.01 μF

Experiment 5

1 EL Digi-designer
1 ±15-V power supply
3 Voltage reference sources (variable, 0–10 or 0–15 V)
1 Quad 2-input NAND gate package (SN 7400)
1 Hex inverter package (SN 7404)
1 Triple 3-input NAND gate package (SN 7410)
1 Bistable data latch (SN 7475)
1 BCD counter (SN 7493)

1 Up/down BCD counter (SN 74193)
1 Voltage comparator (LM 710C)
1 Operational amplifier (SN 72741)
1 Digital-to-analog converter (Datel DAC-9)
1 Pulse generator or clock
1 Volt–ohm multimeter
9 1% resistors: 10 MΩ, 100 kΩ, 50 kΩ, 25 kΩ, 12.5 kΩ, 6.666 kΩ (a combination equal to this), two 2 kΩ, and 1 kΩ
13 Capacitors: 1 μF and twelve 0.01 μF

COMPUTER EXPERIMENTS

In addition to the equipment listed below, this section requires the use of a laboratory computer system, with a general-purpose interface (see Appendices C and F), containing a programmable clock and digital and analog interfaces, including an interface for either an *X–Y* plotter or a storage display oscilloscope.

Experiment 6

1 EL Digi-designer
1 Quad dual-input AND gate package (SN 7408)
1 Quad dual-input NAND gate package (SN 7400)
1 Quad dual-input OR gate package (Signetics SP 384A)
1 Quad dual-input NOR gate package (SN 7402)
1 Hex inverter package (SN 7404)

Experiment 7

1 EL Digi-designer
1 ±15-V power supply
2 Variable power supplies (0 to 15 V)
1 Voltage comparator (LM 710C)
1 Operational amplifier (SN 72741)
1 Digital-to-analog converter (Datel DAC-9)

Experiment 8

Laboratory computer system only

Experiment 9

1 EL Digi-designer
1 Hex inverter package (SN 7404)

1 Resistance substitution box
1 Capacitance substitution box

Experiment 10

1 EL Digi-designer
1 Function generator (noisy triangular wave portion)

Experiment 11

1 EL Digi-designer
1 ±15-V power supply
1 Hex inverter package (SN 7404)
4 Field-effect-transistor switches (CAG-30)
2 BCD counters (SN 7490)
2 Operational amplifiers (SN 72741)
2 0.01-μF capacitors
1 Pulse generator (variable rate)
1 Dual-trace oscilloscope
1 Dual-function generator (10-Hz triangular and square waveforms)

Experiment 12

1 Variable sine-wave source (6-600 Hz)
1 Audio amplifier and speaker (100–10,000 Hz)
1 Dual-trace oscilloscope
2 Resistance substitution boxes
2 Capacitance substitution boxes

Experiment 13

1 Vacuum photodiode in lighttight housing. (This should be constructed with a provision for illumination of the photocathode. It can be made from parts such as an RCA 929 tube and a lighttight box painted black.)
1 Light source (such as mercury arc)
1 Set of optical filters, preferably narrow-band filters
1 Electrometer (such as a Keithley 911)

Experiment 14

1 EL Digi-designer
1 ±15-V power supply
3 Operational amplifiers (SN 72741)
2 Variable resistance boxes

2 1-MΩ 5% resistors
6 Capacitors: two 0.1 μF, two 0.22 μF, two 0.47 μF
1 Oscilloscope
1 pH meter with terminals provided for output of the measured voltage
1 Magnetic stirrer and stirring bar
1 Stepper motor syringe driver (see Appendix E)
Chemicals: Samples of weak and strong acids and bases (e.g., HCl, NaOH, etc.).
Titrants should be approximately 1.0 *M*, while the unknown should be 0.01 to 0.10 *M*.

Experiment 15

1 EL Digi-designer
1 ±15-V power supply
5 Operational amplifiers (SN 72741)
1 Variable resistance box
8 Resistors: five 50 kΩ, three 1 MΩ
3 Capacitors: 0.1 μF
1 Gas Chromatograph-Aerograph 202 with one 10-ft 20% SF-96 column and one 6-ft 20% Carbowax 30-mesh column, and a 1-mV recorder (or the like)
1 Microswitch for GC inlet, or foot pedal or push-button switch
Chemicals: *n*-pentane, *n*-hexane, *n*-heptane, *n*-octane, *n*-nonane, methanol

Experiment 16

1 EL Digi-designer
1 ±15-V power supply
2 Operational amplifiers (one SN 72741 and one Analog Devices P501A)
9 Resistors: 8.2 MΩ, 2.2 MΩ, 1 MΩ, 14.7 kΩ, 10 kΩ, 1 kΩ, 909 Ω, 300 Ω, and 10 Ω
1 Variable resistor (5 MΩ)
7 Capacitors: 100 pF, five 0.01 μF, and 1 μF
1 Diode (1N914)
1 Cell (see Appendix E)
1 Magnetic stirrer and stirring bar
1 2.5-ml syringe
Chemicals: 82 g of potassium dihydrogen phosphate; 42 g of potassium monohydrogen phosphate; 13 g ammonium molybdate $[(NH_4)_6Mo_7O_{24} \cdot 4H_2O]$; 83 g of potassium iodide; 0.06 g of glucose oxidase; 1.0 g of glucose

Experiment 17

1 Spectronic 20 modified as described in the experiment, and preassembled by the instructor. Parts needed for modification include: one 4-rpm synchronous motor (model AR-DA, Hurst Co., Princeton, Indiana, or the like); one reed relay (Magnacraft W105MPCX); one relay (Potter and Brumfield KA5DY); five resistors (1 kΩ, 2 kΩ, 100 kΩ, 1 MΩ, and 69.9 MΩ); two capacitors (5000 pF and 0.5 μF); one transistor (Fairchild 2N3965); two operational amplifiers (SN 72741)

1 ±15-V power supply

2 10-ml Mohr pipets

1 Optical cell for Spectronic 20

Assorted beakers

Chemicals: 0.10 *M* nickel sulfate and 0.10 *M* ethylenediamine

Appendix E

Schematics of Special Devices Used in This Book

GENERAL CONSTRUCTION DETAILS

With one exception, circuit layout and parts placement are not critical for any of the circuits involved. The exception is that, when they are required in the circuit, the 0.01-μF ceramic power-supply decoupling capacitors *must* be placed between the power-supply connections and ground as close to each logic device and operational amplifier as possible. This will prevent noise transients and oscillations from occurring as a result of power-supply-wiring inductances. In addition, reasonable care should be taken to prevent internal ground loops in the devices. Power-supply requirements are fairly noncritical. The supplies should, however, be regulated and have current capacities on the order of about 100 mA for the $\pm$15-V unit, and about 200 mA for the +5-V unit. Inexpensive epoxy-encapsulated units such as Computer Products (Fort Lauderdale, Fla.) Models PM529 (for +5-V unit) and PM552 (for $\pm$15-V unit) can be used.

SINE-WAVE GENERATOR

Several experiments call for a variable frequency, variable amplitude sine-wave generator. If a commercially built circuit is not available, the circuit shown in Fig. E-1 can be used. This circuit is a modified amplitude-controlled Wien bridge oscillator with a buffered output. Two trim-pot adjustments are available for this generator. R_B (50 K) is adjusted for oscillation and a general range of amplitude, while R_A (5 K) is used to trim the amplitude to the desired magnitude. The two adjustments interact, and R_B should be adjusted first.

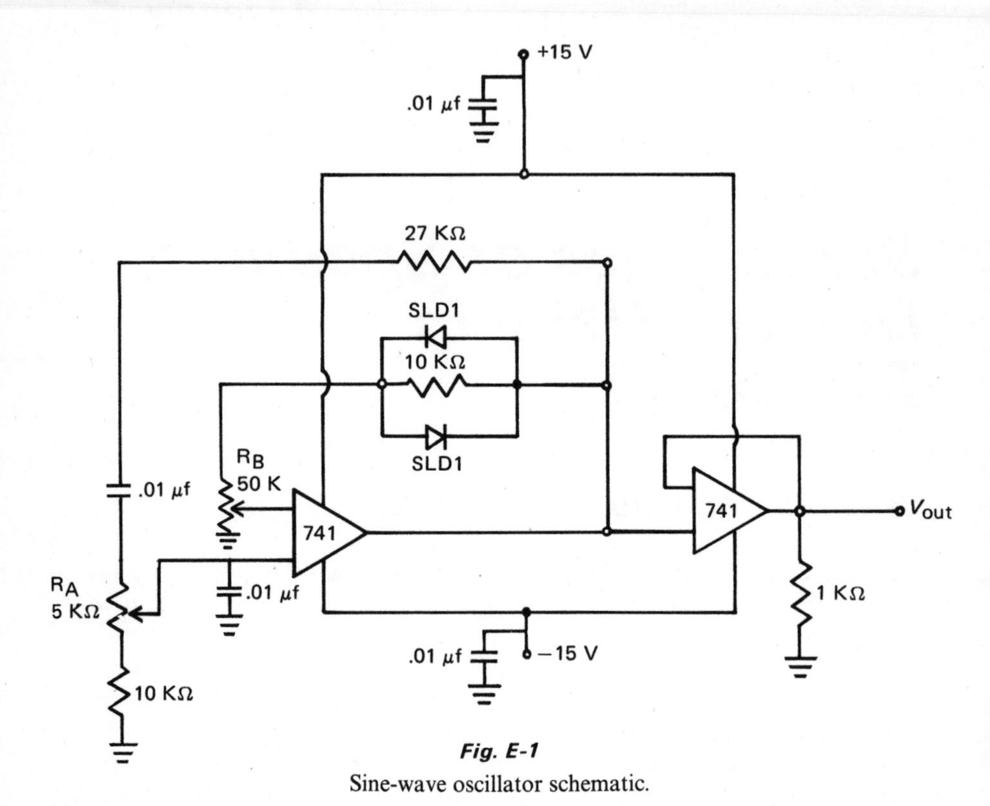

Fig. E-1
Sine-wave oscillator schematic.

EXPONENTIAL-DECAY-FUNCTION GENERATOR

An exponential-function generator is required in Experiment 9 and is optionally necessary in Experiments 10 and 11. We have found that the circuit shown in Fig. 9-2 is adequate for Experiment 9 but has several drawbacks which preclude its use in Experiments 10 and 11. First, there is no provision for adding random (or pseudo-random) noise to the signal. Second, if signal averaging is attempted, the user must wait as long for the capacitor to charge as for it to discharge. For large numbers of scans and long time constants, this measurably increases the time necessary to perform an experiment.

The circuit shown in Fig. E-2 circumvents some of these disadvantages. The exponential generator consists of two FET analog switches (see Experiment 4 for a discussion of FET's) which alternately charge and discharge a timing capacitor. The 100-μF capacitor (C_1) is chosen to give a discharge with a time constant between 0.1 and 1.1 sec depending on the value of the variable resistor R_A. The charging-time constant is determined by C_1, R_B, and the series resistance of FET switch 1; for the values shown here, the

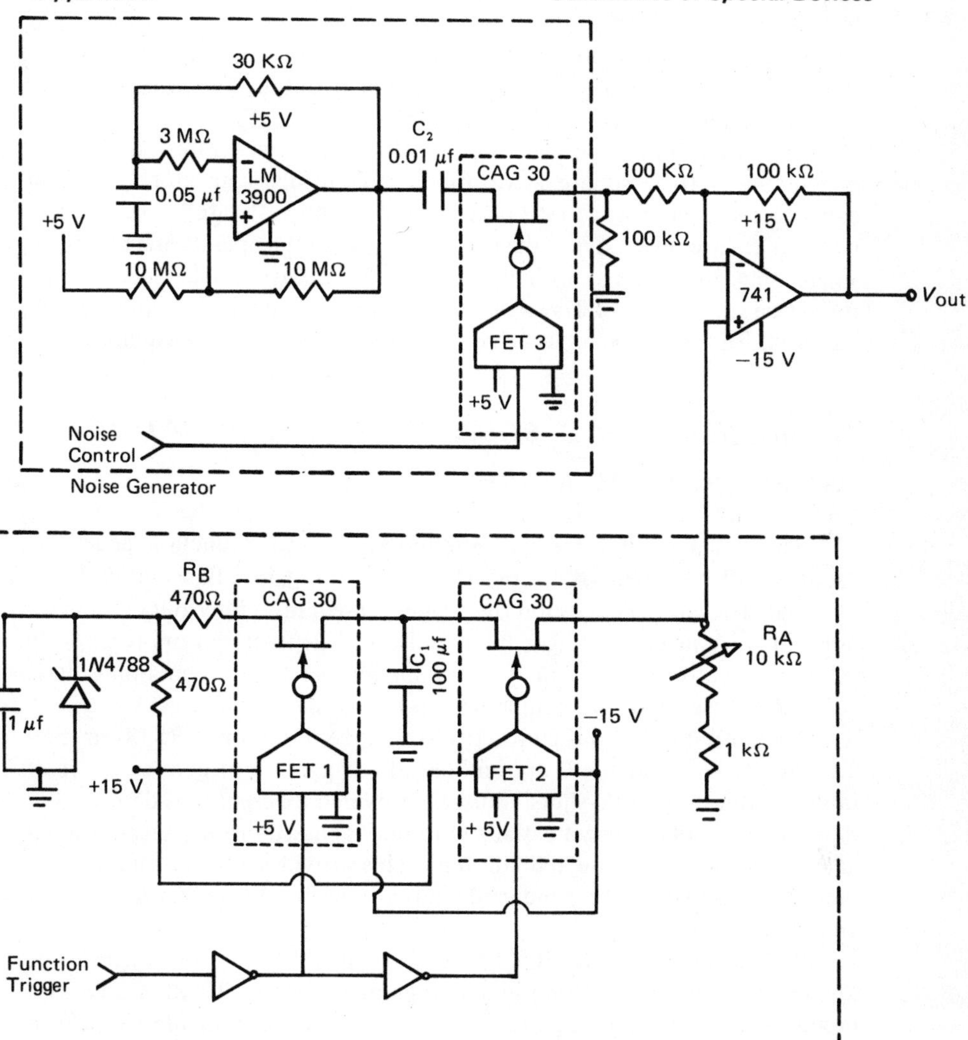

Fig. E-2
Exponential-waveform generator with optional noise.

charging-time constant is approximately 0.05 sec, providing for a quicker reset condition. The output signal is approximately 8 V in magnitude, depending on the Zener diode used.

When the input gate signal is low, switch 1 is on, permitting C_1 to charge through R_B and the R_{ON} of the switch to the potential level established by the Zener diode. On the alternate half-cycle, when the input gate signal is high, switch 2 is on, and C_1 discharges through R_A and R_{ON} of the switch. The gate-signal duration should be long enough to allow the capacitor to completely discharge (roughly five time constants) on output; the recharging time is approximately 0.3 sec for the components shown in Fig. E-2.

The "random" noise generator shown actually generates an approximately 5-V, 400-Hz square wave; this is fed through capacitor C_2, which converts the square wave into positive and negative "noise" spikes. These noise spikes are then fed into the inverting input of the operational amplifier, where they are summed with the exponential decay signal. Noise can be turned on and off by FET switch 3. (If computer control of the noise is not required, FET 3 can be replaced by an ordinary mechanical switch.)

TRIANGULAR- AND SQUARE-WAVE GENERATOR (WITH OPTIONAL NOISE)

The schematic for the triangle- and square-wave-function generator, required in Experiments 4, 10, and 11, is given in Fig. E-3. IC 9 and IC 10 are optional. They act to condition the trigger pulse and to filter out extraneous noise in it. A negative trigger input pulse will generate an approximately 0.2-msec negative pulse from IC 10. Function generators containing these IC's can be triggered by a simple RS flip-flop or by a bounceless switch.

The output of IC 10 is connected to the reset input of an RS (Reset–Set) flip-flop. (If IC 9 and IC 10 are omitted, the trigger input goes to the reset input; in this case, "one-shots" must be provided *externally*, as described in Experiment 10.) The output of the RS flip-flop is passed through an additional gate to provide better rise and fall times. This provides a better square-wave output. The square wave is buffered (and optionally adjusted for dc level) by amplifier IC 15.

Operational amplifier IC 11 is used as an integrator to integrate the square-wave output, resulting in a triangular waveform. IC 12 is used as a noninverting summing amplifier in order to offset the triangular waveform completely above ground. The offset is necessary to provide a correct set voltage level for the RS flip-flop. To provide for a symmetrical waveform, the trim pot connected to the noninverting input of IC 11 is adjusted while triggering the RS flip-flop.

The waveform-generator circuit operates as follows: When a negative-going trigger signal is applied to the trigger input, the output of the RS flip-flop goes from +5 to 0 V. Because this output is inverted, the input to IC 11 becomes +5 V, and this amplifier begins integrating the 0 to +5 V step function. This produces a negative ramp. The negative ramp is offset by IC 12 and the noninverting input of IC 11, so that it goes from a positive voltage towards zero. As it approaches zero, it resets the flip-flop through the diode.

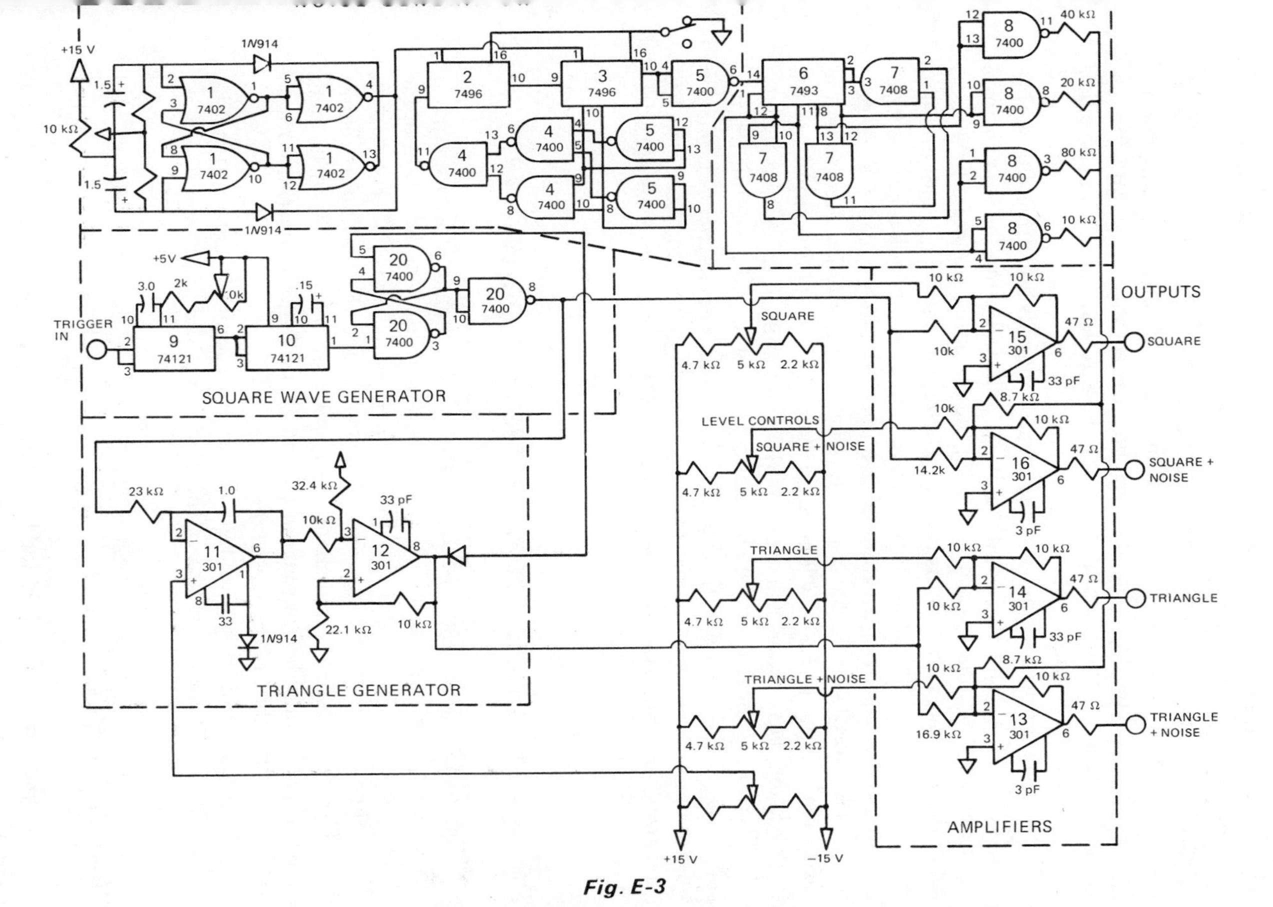

Fig. E-3

Triangular- and square-wave generator (with optional noise). Capacitors in microfarads; resistors in Ohms, $\frac{1}{2}$ watt. All amplifiers decoupled with 0.01 μF capacitors on pins 4 and 7.

The inverted output of the flip-flop then drops to 0 V. The integrator IC 11 then produces a ramp going from a negative voltage towards zero. The integrator output reaches 0 V and remains there until another trigger pulse is applied. With the component values shown, a full-duty cycle waveform is produced in about 100 msec. By changing the input resistor and bypass-capacitor values of this integrator (IC 11), the waveform period can be either lengthened or shortened.

The triangle- and square-wave-generator outputs are applied to amplifiers IC 13, 14, 15, and 16. Amplifiers IC 13 and IC 16 sum the triangle- and square-wave functions, respectively, with the output of the noise generator. All of the amplifiers are used as inverting summing amplifiers, with optional dc level control. With the input resistances specified, the noise-free triangle- and square-wave output magnitudes are approximately 6 and 5 V, respectively. The noisy signals have slightly smaller magnitudes. Reducing the values of the input resistors will increase the magnitude of the output signals, while raising these values will decrease the magnitude of the output signals.

The noise generator is a slight modification of the random-shift register and noise oscillator used in the Varian XL-100 nuclear-magnetic-resonance spectrometer. It is also possible to use several high-frequency coupled RC oscillators, but with less desirable results, since the noise they generate is less "random."

PROGRAMMABLE STEPPER MOTOR DRIVER

Figure E-4 shows the schematic for the programmable stepper motor driver needed for Experiment 14. This circuit is a slight modification of the circuit the Heath Corporation uses for their Model IR-18 $X-Y$ recorder. The trailing edge of each high-input pulse causes the JK flip-flop to change state. The normal and complementary outputs are then amplified through Q_2 and Q_3; the amplified outputs of Q_2 and Q_3 drive the stepper motor one notch each time the JK changes states.

The syringe holder and plunger driver are assembled as shown in Fig. E-5. Figure 14-3 shows a picture of the final assembled apparatus.

REACTION CELL

The reaction cell required for Experiment 16 is shown in Fig. E-6. The reaction takes place in a small compartment of about 5 ml total capacity, immersed in a thermostated liquid for temperature control. The electrodes are prepared by sealing platinum wire (22 gauge) into $\frac{1}{8}$-in. glass tubing. The

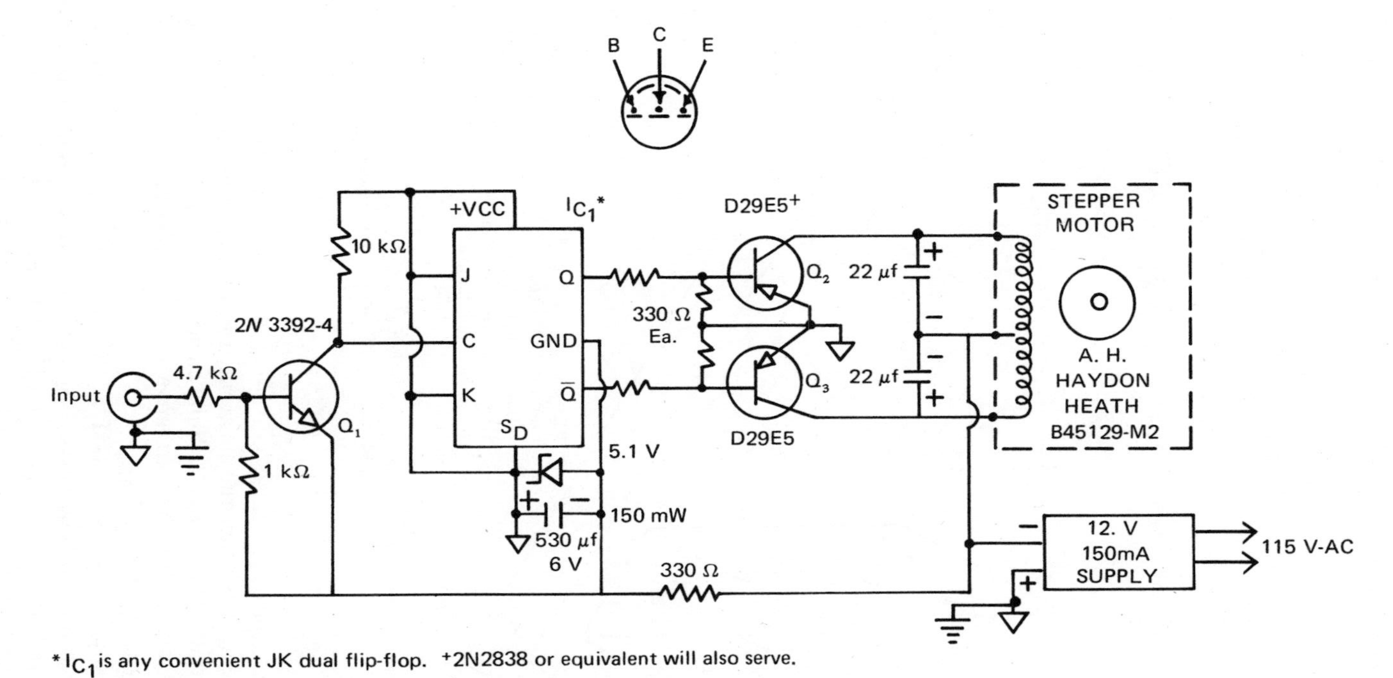

* IC_1 is any convenient JK dual flip-flop. +2N2838 or equivalent will also serve.

Fig. E-4
Stepper-motor-driver schematic.

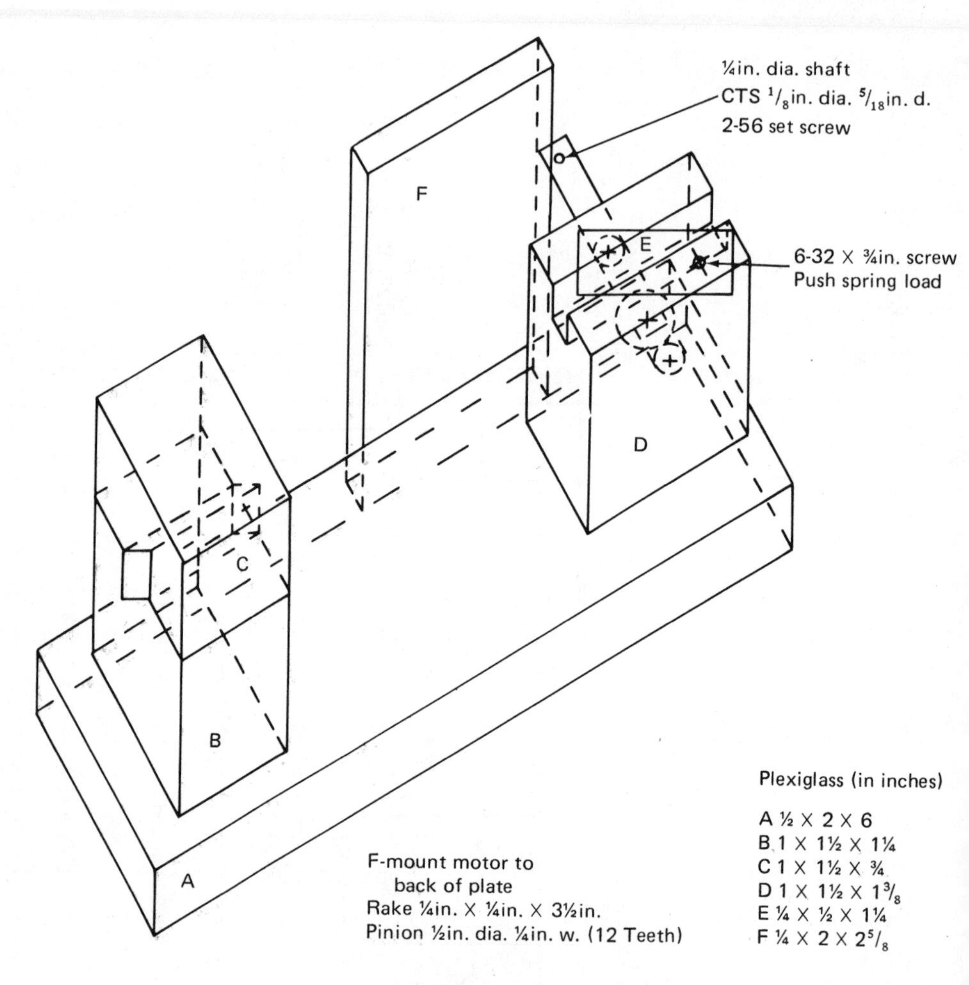

Fig. E-5
Syringe-plunger-driver mechanical assembly.

rotating electrode is sealed onto $\frac{1}{4}$-in. tubing to fit the stirring chuck. This seal must be made so that the electrode rotates smoothly and does not wobble. An alternate procedure is to build an adapter for the $\frac{1}{8}$-in. tubing and stirring chuck. Contact with the platinum wires is made by a copper wire dipping into mercury in the electrodes.

The electrode is rotated at about 1800 rpm and provides efficient stirring in the cell. The Sargent Model E Electrotitrator stand serves as the stirring motor (E. H. Sargent and Co., Chicago, Ill.). The cone-drive motor

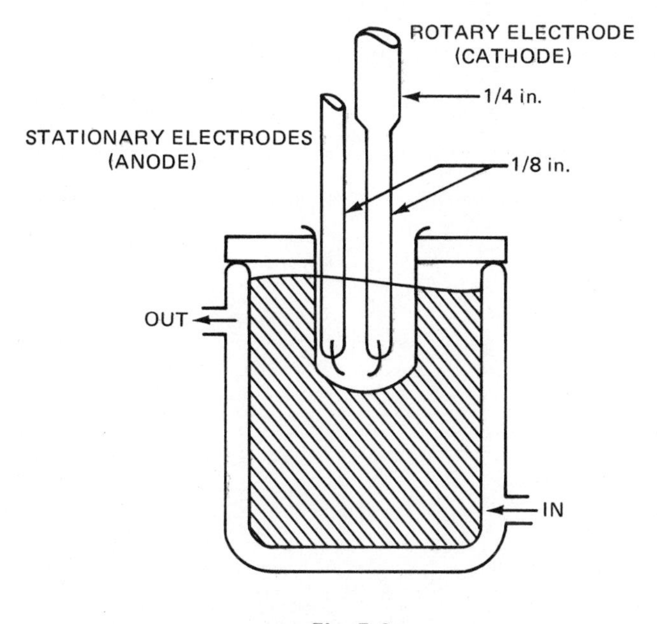

Fig. E-6
Thermostated reaction cell.

available from this supplier works equally well (operated at maximum stirring rate) and offers the advantages of economy and variable stirring rate. The 600-rpm synchronous rotator is not satisfactory for this experiment. The reader should consult ref. E-9 for further details.

Appendix F

*Detailed Description of Data-Acquisition Interface for Hewlett-Packard 2100-Family Computer Systems**

Some minicomputer manufacturers also offer general-purpose laboratory data-acquisition-interface hardware as an optional accessory. Typical examples are the LAB-8 and LAB-11 systems offered by the Digital Equipment Corporation. However, if that option is not available, or if the available hardware is not completely suitable for the *educational environment*, it may be necessary to construct the interface for your particular minicomputer system. This was the case for the laboratory computer system used at Purdue. In this Appendix we will describe the design philosophy and detailed characteristics of the laboratory interface developed for the Hewlett-Packard 2100-family (models 2116, 2115, 2114, 2100, and 21MX) of minicomputers.

DESIGN PHILOSOPHY

The basic approach taken was to provide an easily accessible interface panel which students could interpret functionally and use primarily with patchcord connections. The functional features of the interface panel included:

Analog-to-digital conversion (11-bit, ± 10 V, 25 μsec, preceded optionally by a track-and-hold amplifier)

*Prepared by E. D. Schmidlin, Chemical Instrumentation Group, Chemistry Department, Purdue University, Lafayette, Indiana.

Digital-to-analog conversion (16-bits, ± 10 V)
Four-channel multiplexing (using 4 separate analog FET switches)
Two command bits (ENCODE)
Two status bits (FLAG)
Two 16-bit digital input/output channels
Power, ground, and indicator lamps
Programmable clock (0.01 Hz to 100 kHz)
Various logic functions (AND, OR, NOT gates; flip-flops; one-shots level converters, switches, etc.)
Two slots for special plug-in printed circuit cards

In total, three separate input/output channels are used for the general-purpose laboratory interface: one for the DAC; one for the ADC, programmable clock, Command/Status, and 16-bit digital I/O bits; and one for separate 16-bit digital I/O and Command/Status bits. The external connectors are rack-mounted as shown in Fig. F-1. Figure F-2 illustrates a suggested wiring diagram for a typical application.

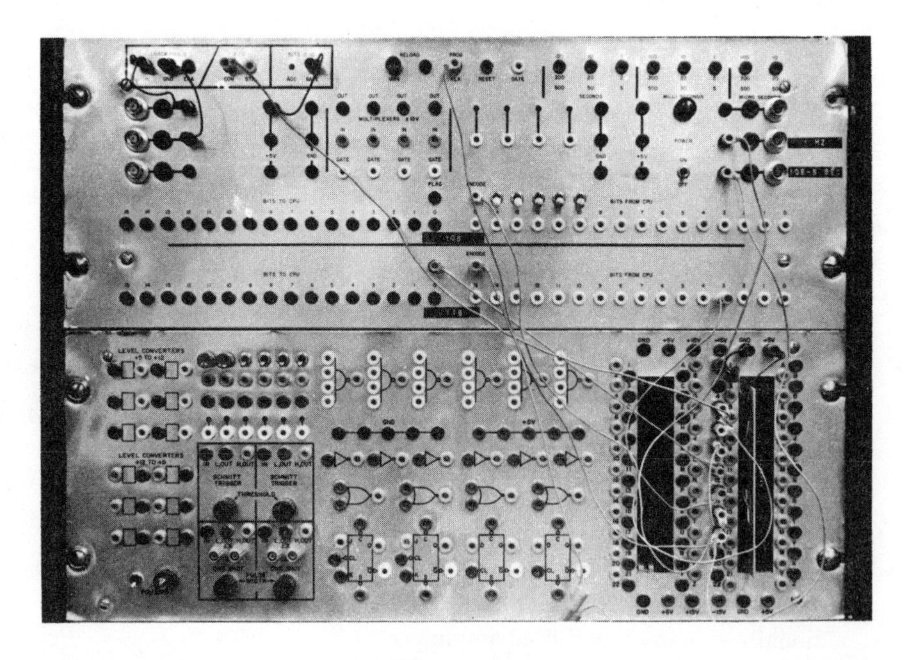

Fig. F-1

The Purdue interface module for Hewlett–Packard 2100–family computer systems.

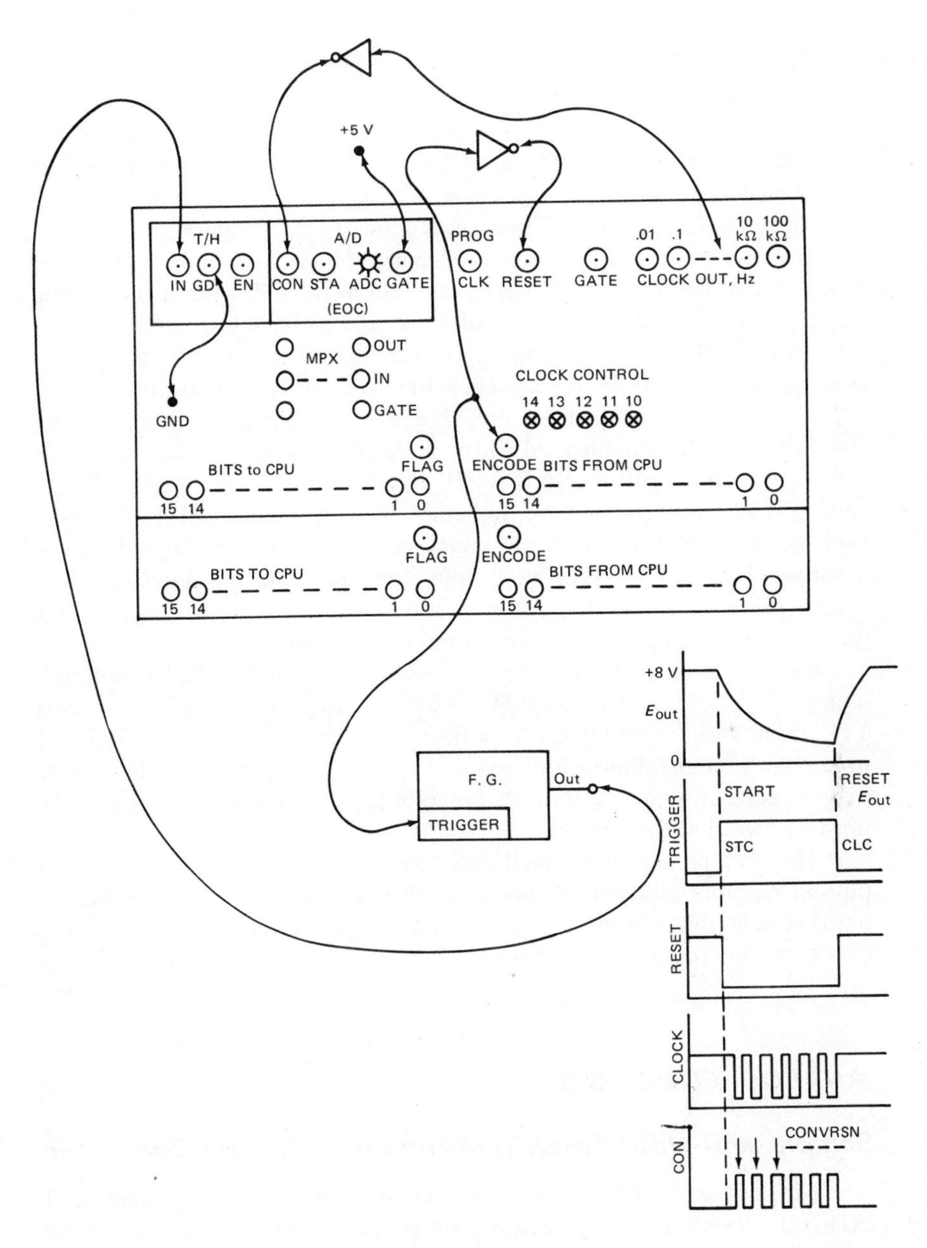

Fig. F-2

Wiring diagram for Experiment 9.

DESIGN DETAILS

The interface is basically divided into three areas of concern, namely, the analog section, the digital section, and the logic section. The analog section contains facilities for obtaining data, converting the data, and entering it into the computer. It is comprised of a sample-and-hold (S/H) amplifier and an analog-to-digital converter (ADC). Also included are four analog switches.

The digital portion of the interface is concerned with digital communication and/or control. It is comprised of several functions. They are indicated below and will be explained in greater detail in the following sections. The functions are: S/H and A/D control, FLAG and ENCODE line, BITS TO CPU, BITS FROM CPU, digital-to-analog conversion (DAC), BITS 0–10 OPTION control, PROGRAMMABLE CLOCK, CLOCK DECADES, and BIT MONITORS.

The above-mentioned features are associated with 3 different I/O channels and are implemented through the H. P. general-purpose duplex register interface cards (Model No. H. P. 02116-6195), 12-V negative-logic version. The logic functions are implemented with standard MSI TTL logic packages mounted on a separate panel. Available functions include six 4-input NAND gates, six INVERT gates, four NOR gates, four flip-flops, two one-shots, two Schmitt triggers, six level converters, six buffered switches, six indicator lamps, and two slots for 44-pin plug-in cards.

For the most part, the interface is comprised of TTL logic, where a logical "1" is 3.5 V and a logical "0" is 0 V. Although most of the circuitry is TTL, input and output buffers have been used throughout the interface. All inputs are emitter-follower buffered and reside at a logical "0." This means that to make an input perform its function (enable or assert), a logical "1" must be placed at the terminal.

The outputs are all buffered and have open collector transistors with pull-up resistors attached. This means that little if any damage will take place should two outputs be accidentally or intentionally tied together. Clock outputs should *not* be tied together.

ANALOG FUNCTIONS

Sample-and-Hold Amplifier–Analog-to-Digital Converter

The sample-and-hold (S/H) amplifier and the analog-to-digital converter (ADC) used within the data-acquisition portion of this interface are commercially available units. A SHA-1A sample-and-hold amplifier and an ADC-Q analog-to-digital converter available from Analog Devices Inc.

were selected. Complete specifications for these two units can be obtained from the manufacturer's data sheets.

The front panel analog input marked TRACK/HOLD IN is internally connected to the sample-and-hold amplifier which in turn is directly connected to the analog-to-digital converter. The sample-and-hold amplifier operates in a track-and-hold fashion and is connected to operate normally in the track mode. If the hold feature is desired, the ENable terminal must be made a logical "1," and the holding action will be properly initiated when conversion is desired.

Conversion by the ADC is initiated by an input to the CONvert terminal, for example, by the output from the programmable clock or one of the decades. The conversion is initiated by the return to "0," that is, the transition from a logical "1" to a logical "0." This choice of gating is further discussed in the "Implementation" section of this appendix.

An end-of-conversion (EOC) pulse is available at the front panel terminal and is also used by the interface to generate a FLAG bit when the ADC output is passed on to the computer.

The analog input range is set for ± 10.23 V, thereby permitting a resolution of 10 mV per bit from the ADC.

Analog Switches

Included as a part of the data-acquisition section are four analog switches. These switches will pass an analog signal when a logical "1" is placed on their gate input. Each switch operates independently and will pass a signal of ± 10 V. All signal sources should be of low impedance (less than 100 Ω).

Digital-to-Analog Converter

The DAC is a H. P. model J04-6933A, 16-bit ± 10-V device interfaced through I/O channel 16_8. A standard duplex register interface card is used as specified above. This DAC provides a 20-μsec conversion time with ± 0.5 mV resolution.

DIGITAL FUNCTIONS

Bits from CPU

Sixteen data bits from the computer via I/O channel 10_8 are buffered and made available at the front panel with positive true logic. These points may be used to drive other circuitry. If the programmable clock is used, bit 15 is reserved as a strobe line to change the clock rate. The clock rate is established via information contained in bits 10 to 14 at the time the strobe occurs.

Bits to CPU

Sixteen jacks are available on the front panel to enable the simultaneous transfer of 16 bits of information from the interface into the CPU through I/O channel 10_8. These inputs are positive logic and reside at logical "0" with no inputs attached.

FLAG and ENCODE

The FLAG and ENCODE bits are used by the computer as outlined in the *Pocket Guide to Interfacing H. P. Computers* (Hewlett-Packard Co., Cupertino, California). The FLAG signals the computer with input information while the ENCODE signals the interface with output information. Both lines work as a strobe and function at the interface with positive true logic.

Bits 0–10 Control

Data bits to the computer can be entered from the front panel jacks as mentioned above or directly from the ADC outputs. The special control GATE on the front panel when ON (logical "1") will cause the output of the ADC to be placed on bits 0 to 10, with bits 11 to 15 the same as bit 10 to extend the sign bit for 2's complement notation. A special front panel indicator lamp indicates when the ADC outputs are enabled. Bit 9 is the most significant bit (MSB), while bit 0 is the least significant bit (LSB).

Sample/Hold and Analog-to-Digital Control

A minimum "track" period is hard-wired into the interface as recommended by the manufacturer of the S/H and ADC. This period provides sufficient settling time for the S/H when the ADC is programmed to run at the maximum rate and is initiated by a CONvert command at the front panel, which is a transition from a logical "1" to a logical "0." Following this (~ 5 μsec) track period, an ADC conversion is initiated, and the S/H is set to the "hold" mode simultaneously.

A buffered end-of-conversion (EOC) signal is brought out to the front panel for external use; this signal is a logical "0" during the time the ADC is busy.

Clock and Decades

The clock of the interface is a 20-MHz oscillator, Motorola K1091A. The output from this clock is reduced to 1 MHz and divided down via a decade register; the outputs of each decade are available at the front panel terminals. The rate multiplier is established by output bits 13 and 14 as discussed in the next section. The decades can be reset to zero or held at zero by applying a logical "1" at the RESET terminal.

Programmable Clock

The programmable clock is controlled via the contents of bits 10 to 14 as seen by the programmable clock. Bits 13 and 14 set the basic interval of both the programmable clock and the decades, while bits 10, 11, and 12 establish the decade interval of the programmable clock. Table F-1 shows the programming structure.

The programmable clock will change to a new interval only upon completing an existing interval or when forced by a logical "1" at the RELOAD terminal. The PC GATE input will inhibit the PROG CLOCK output when a logical "1" is present.

A special feature of the programmable clock is the front-panel switch control which when used overrides the computer-controlled clock rate. If one or more of the data switches located on the front panel directly above bits 10 to 14 are switched on (UP), the clock rate is established by the switch register. An up switch indicates a logical "1".

Bit Monitors

Four lamp indicating monitors marked AUX1, AUX2, AUX3, and AUX4 are provided on the front panel. A logical "1" when placed at the input will cause the lamp to light.

Bits from CPU (Special)

The 16 data bits and ENABLE from I/O channel 11_8 are also brought out to the front panel. All bits are buffered and none are dedicated.

Table F-1

Programmed Clock Intervals

Basic interval	14	13	12	11	10	Decade interval
÷1	0	0	0	0	0	10 μsec
÷2	0	1	0	0	1	100 μsec
÷5	1	0	0	1	0	1 msec
÷10	1	1	0	1	1	10 msec
			1	0	0	100 msec
			1	0	1	1 sec
			1	1	0	10 sec
			1	1	1	100 sec

Bits to CPU (Special)

The 16 data bits and FLAG can be entered also via I/O channel 11_8 using jacks on the front panel. All inputs are buffered and require a logic "1" at the jacks.

IMPLEMENTATION

General

Figures F-3a and F-3b, along with the discussion below, will aid the reader in better understanding the interface. The operational block diagram (Fig. F-3a) differs somewhat from the hardware block diagram (Fig. F-3b), but as the discussion continues the two will quickly interrelate.

All construction for the interface is housed within a standard card rack for 4.5-in. cards. The front panel is a 7-in. panel, and all features were described in the preceding section.

Sample/Hold and Analog-to-Digital Converter

The Analog Devices, Inc., sample-and-hold amplifier SHA-1A was selected to maintain system compatibility with the analog-to-digital converter model ADC-Q. The ADC-Q, with a 12-bit resolution, was selected for this interface. Only 11 bits of the conversion word are used by the interface which accepts a full ± 10-V analog input signal. The output bipolar code is 2's complement with the most significant bit acting as the sign bit, and the least significant bit is unused.

The most significant bit (sign) from the ADC is associated with bits 10 to 15 of the computer, and each bit follows in descending significance. Therefore, the resolution as used by the interface is 1 part in ± 1023 or approximately 10 mV per bit.

The status output of the ADC is connected to the control input of the S/H (see Fig. F-4) and is used to initiate the hold mode of operation when an ADC conversion is started. The convert command is seen by the ADC as an output from a one-shot shown in Fig. F-5 (pin *Y*). This particular one-shot is less susceptible to power supply noise and was adjusted for a 5-μsec pulse width, which will provide sufficient track and settling time for the S/H between successive conversions.

A special ENable feature for the sample-and-hold has been provided. The ADC status bit will not trigger the hold feature of the S/H amplifier unless the ENable signal is high.

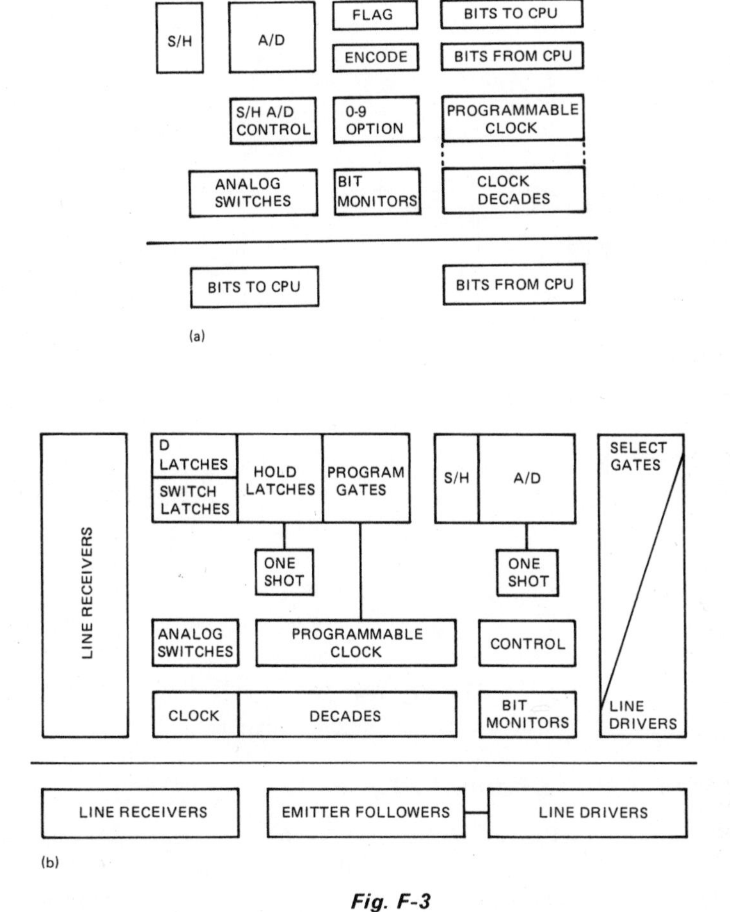

Fig. F-3
(a) Interface operational block diagram. (b) Interface hardware block diagram.

Analog Switches

Four analog switches (Fig. F-5) are provided. These switches use a Teledyne CAG-30 general-purpose FET analog gate. The switch is turned on by a logical "1" applied to the GATE on the front panel, and displays an ON resistance of 60 to 100 Ω. Signals of ± 10 V can be switched.

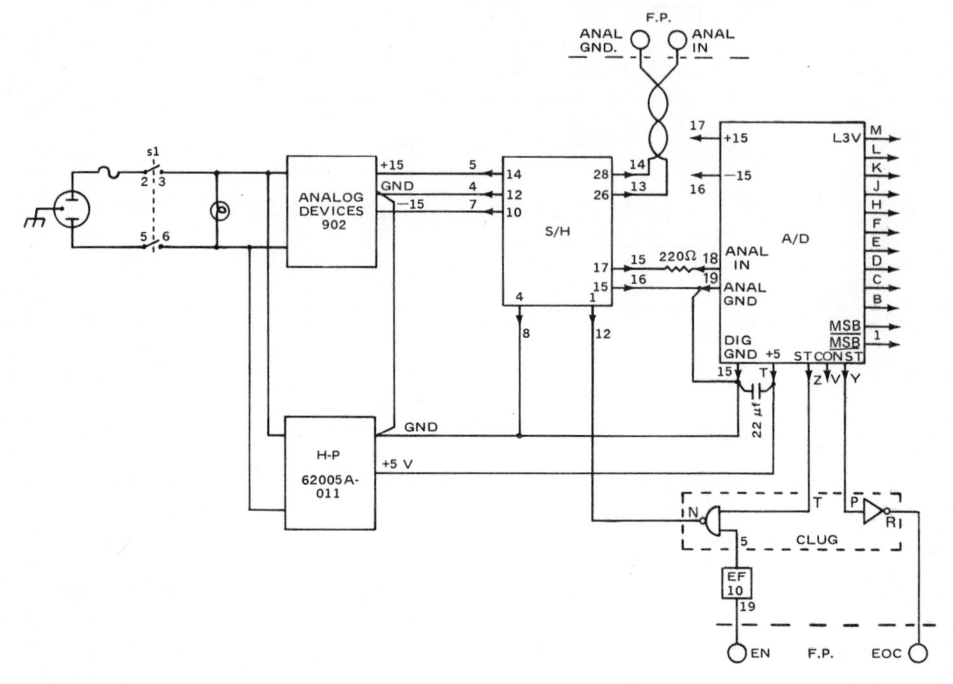

Fig. F-4

Analog system and supply for interface.

Bits from CPU (Channel 10_8)

Bits from the CPU are received by the interface using SN 7405 open-collector inverting buffers. A single 470-Ω load resistor is added to each input line and returned to +5 V, thereby providing 10 mA of current to the I/O driver of the H. P. duplex register.

The open collectors of the inverting buffers are returned to +5 V through 2.2-kΩ resistors. All 16 bits and the ENCODE bit are brought out to the front-panel terminals. Bits 10 to 14 contain programmable clock information when bit 15 is strobed by software control. These bits may be used for other purposes if the programmable clock is not used.

Bits to CPU (Channel 10_8)

Open-collector buffer inverters of the SN 7405 type are used to drive the 16 bits and FLAG line to the duplex register card. Load resistors for the open collectors are located on the duplex cards and provide approximately 10 mA of current.

Inputs for the line drivers consists of grounded input emitter followers.

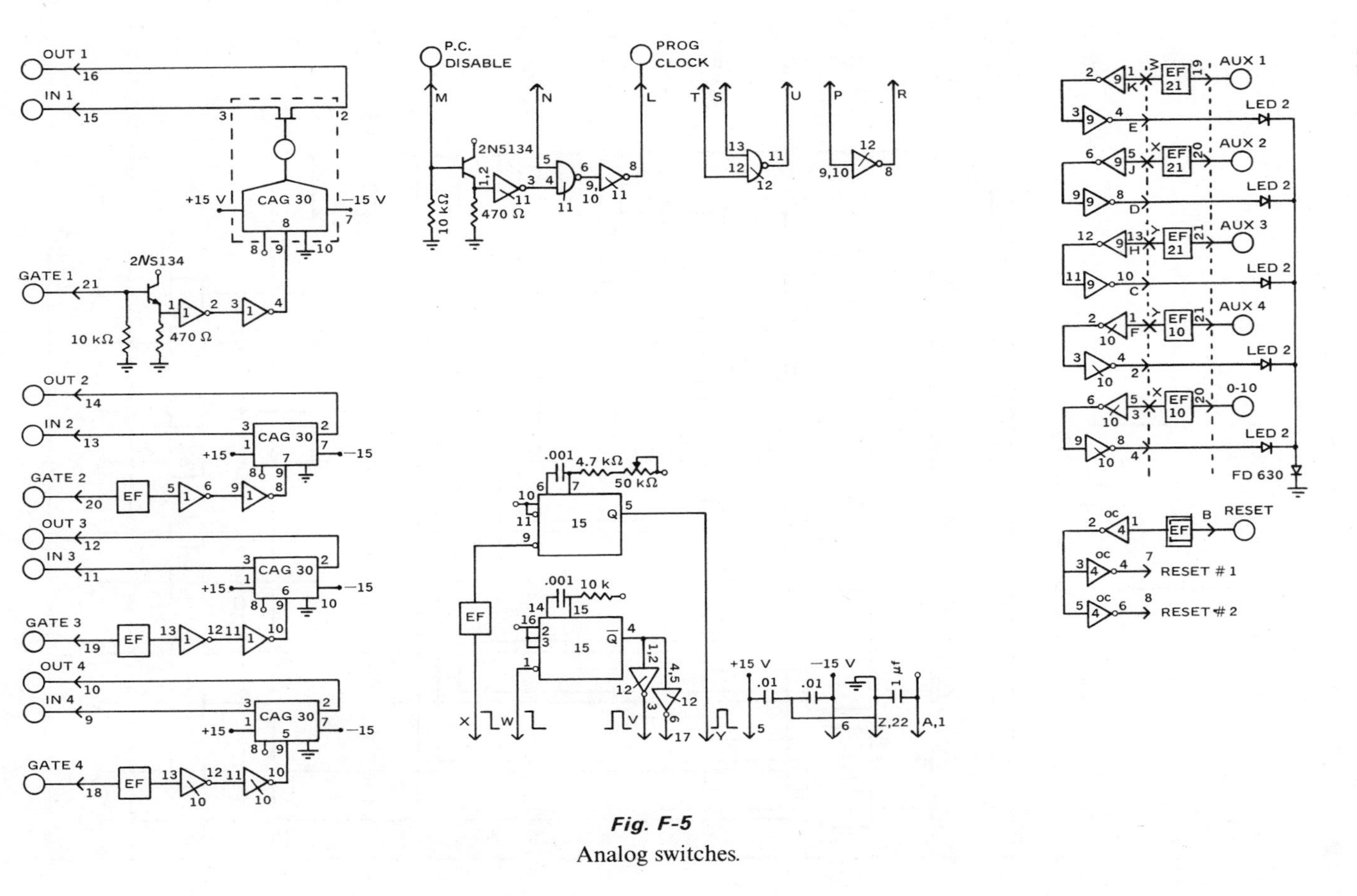

Fig. F-5
Analog switches.

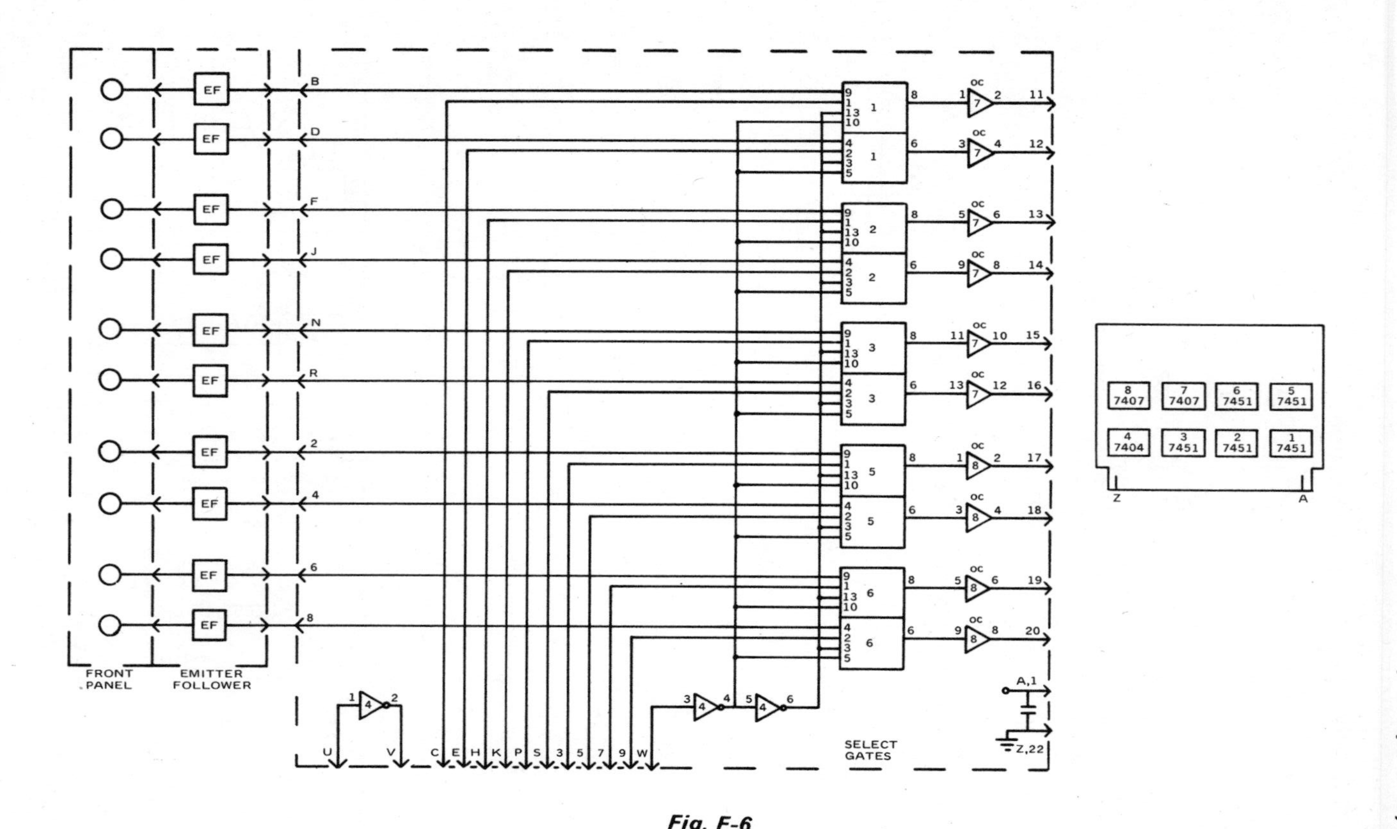

Fig. F-6
Input buffer, select gates, and output drivers.

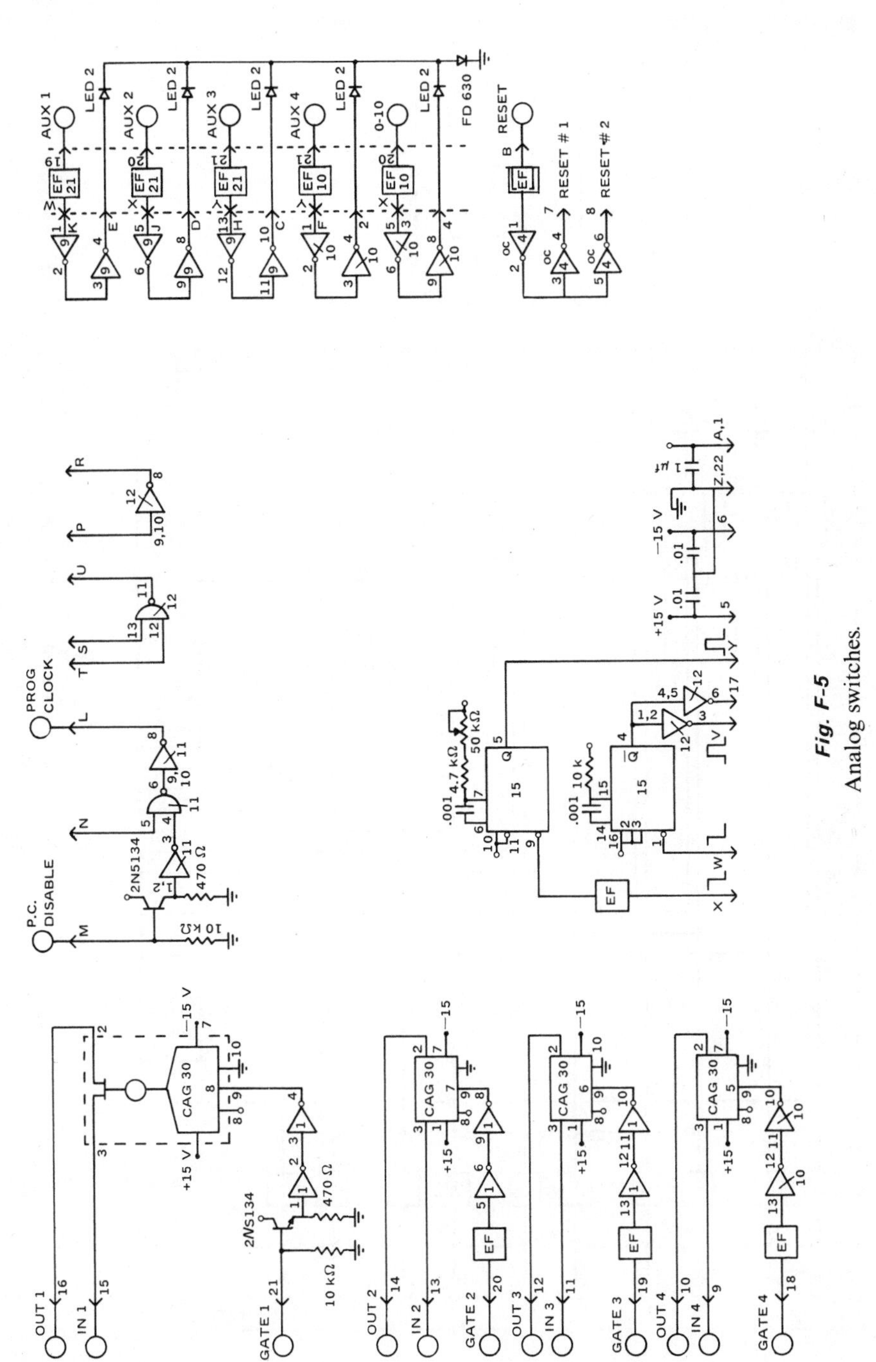

Fig. F-5
Analog switches.

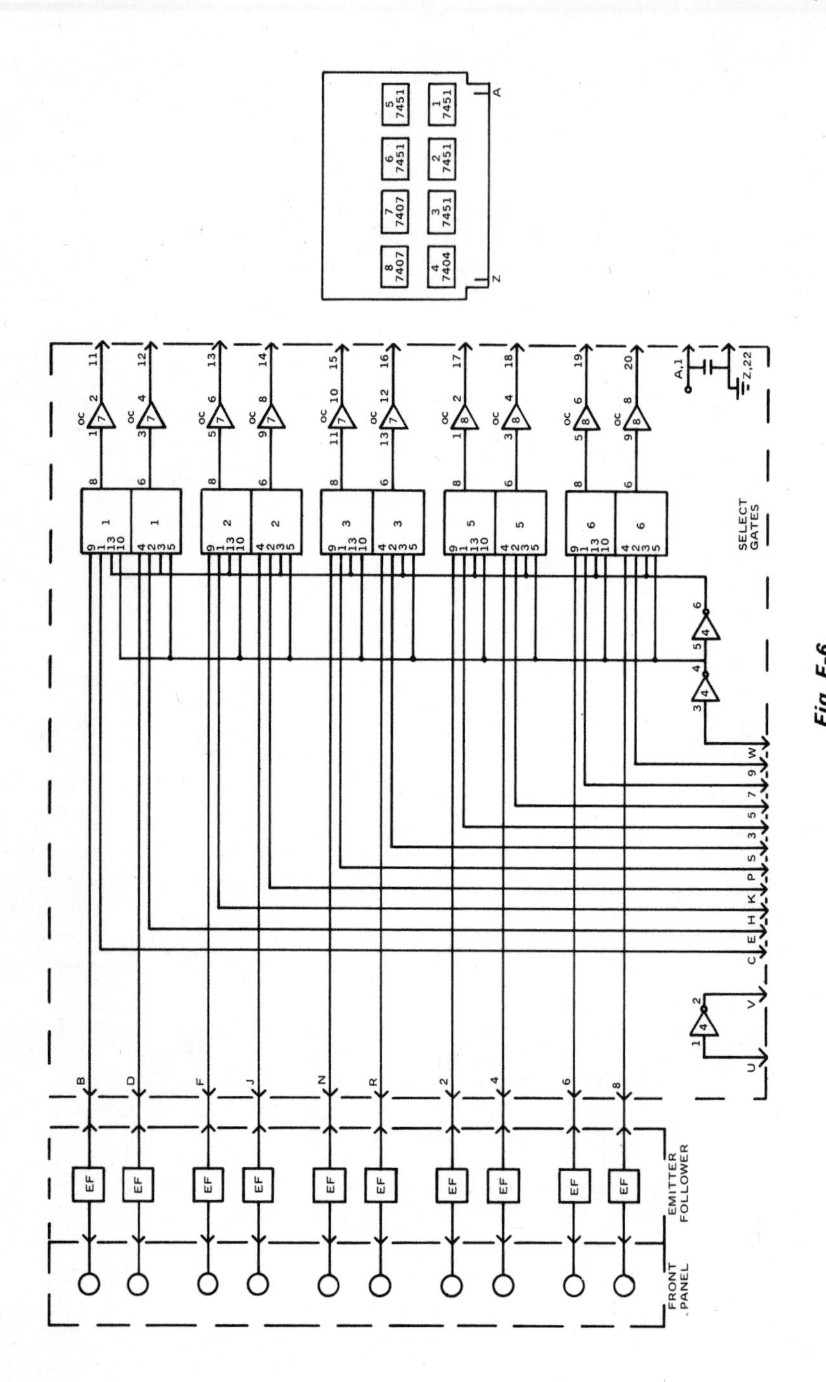

Fig. F-6
Input buffer, select gates, and output drivers.

Select Gates

The information which is sent to the computer on channel 10_8 can be supplied from the ADC or from the front-panel jacks. The select gates are controlled by a common line generated by the 0–10 bit select (GATE on front panel). (Figure F-6 shows the gates for bits 0–9. Another identical card is used for bits 10–15 and the FLAG and ENCODE bits.) The gates are dual 2-input AND-OR INVERT gates of the SN 7451 type. All inputs from the front-panel jacks are emitter-follower coupled to the select gates.

Clock and Decades

A Motorola K1091A (20 MHz) crystal-controlled oscillator provides the basic clock frequency used by the interface (see Fig. F-7). This 20-MHz signal is divided down by 20 at IC15 and IC14. The output from IC14 is passed through a divide by 1, 2, or 5 circuits consisting of IC5, and the outputs are multiplexed by the programmable clock (see below) and passed onto the countdown decade registers consisting of eight SN 7490's connected for symmetrical outputs. Each decade is buffered, and the output is brought out to the front panel.

Programmable Clock

The programmable clock consists of two sections. IC13 is a digital multiplexer which selects one of three inputs, namely 1, 0.5, and 0.2 MHz; whereas IC4 is used to select which decade output will be programmed out and presented to the front panel via the programmable clock gate.

IC11 on Fig. F-5 is the disable clock gate, and with a logical "1" placed at the input, the programmable clock terminal will be held low. This gate will permit removal of clock pulses without resetting the clock register. A special RESET terminal shown in Fig. F-5 will cause the clock register to reset to zero and hold at zero as long as logical "1" input is present.

Latches and Program Gates

Information on bits 10 to 14 from the computer I/O channel 10_8 can be strobed into the interface by using bit 15 from the computer as a strobe. The information is strobed into five D latches. The contents of the switch register along with the information from the D latches are strobed into a second group of D latches. This strobe exists only at the time of a programmed output (interval completion). Complete details for the latches and program gates can be found on Fig. F-8. IC5 senses the switch register and will enable if any switch is placed up. IC11D is used to gate information from the computer when all switches are down (disabled). A special RELOAD function is available when wanting to override the program strobe.

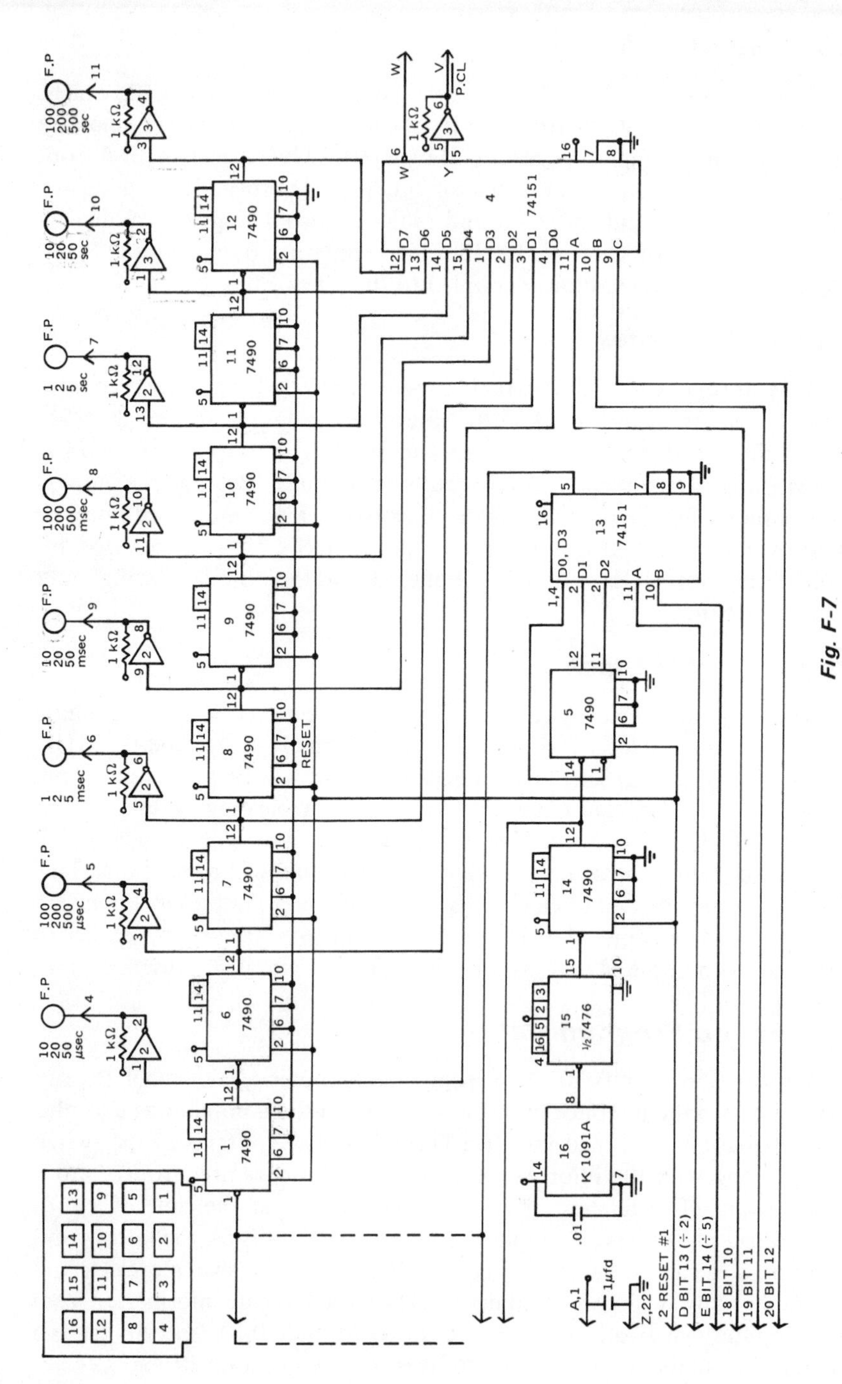

Fig. F-7
Programmable clock for interface.

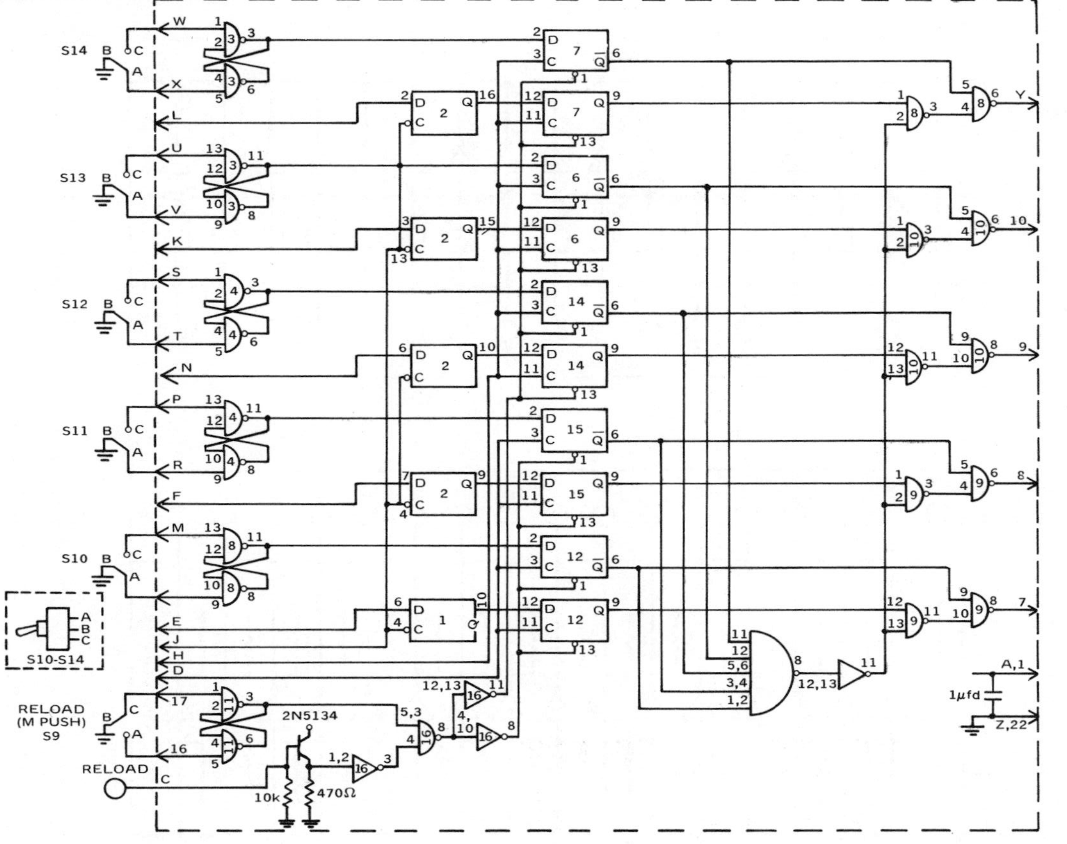

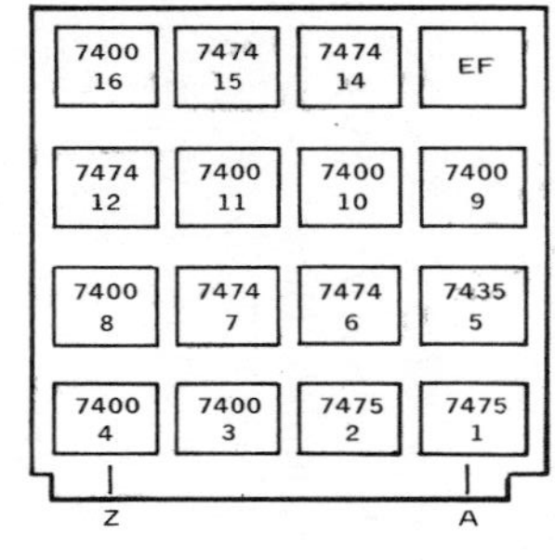

Fig. F-8
Interface program gates and latches.

Bits to and from CPU (I/O Channel 11_8)

A second 16-bit duplex register on channel 11_8 of the CPU is buffered to the interface. Data bits from the CPU are buffered the same as those using I/O channel 10_8 and are present at the front-panel terminals. Data bits to the CPU from the front panel terminals are emitter-follower buffered before being presented to the line drivers. Both receivers and drivers are shown in Fig. F-9.

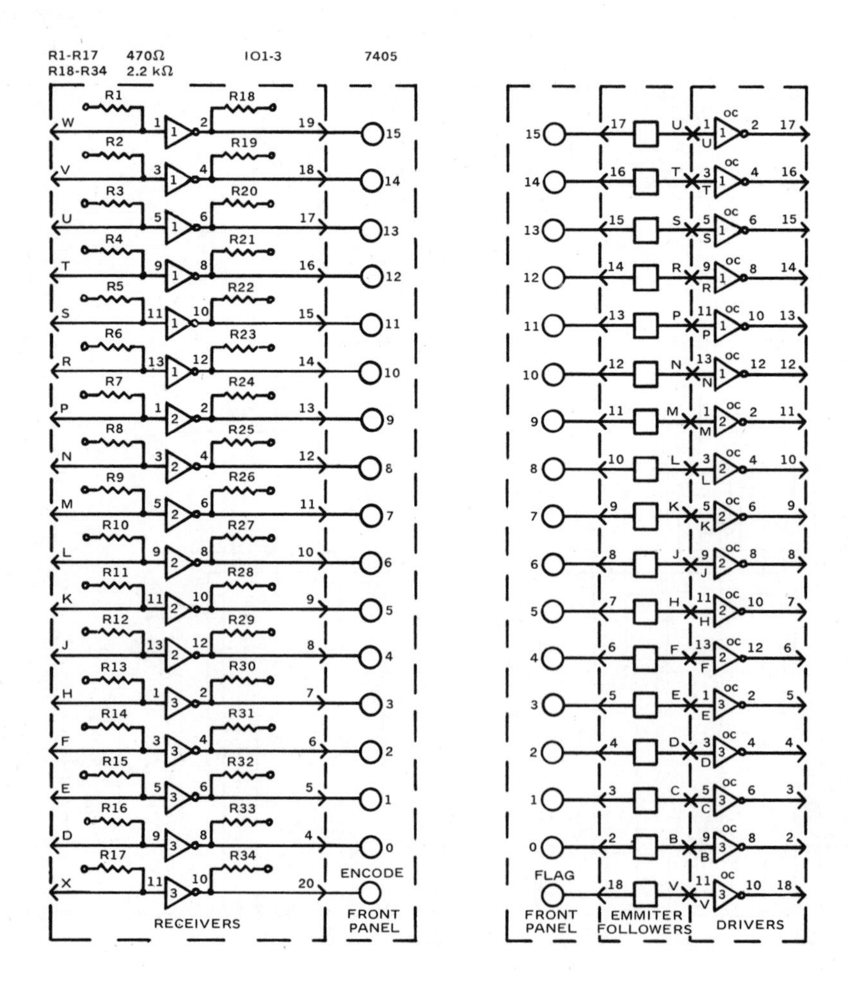

Fig. F-9

Receiver drivers for interface.

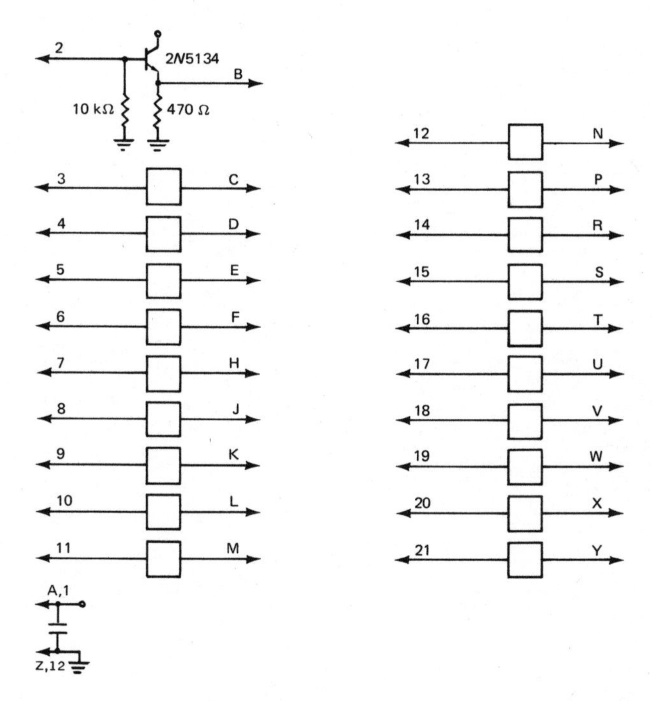

Fig. F-10
Emitter-followers for interface inputs.

Bit Monitors

Four separate bit monitors with lamp outputs are available for either testing or routine use. These units are emitter-follower coupled and require a logical "1" to turn on the lamps. Terminals AUX1 to AUX4 on the front panel are used as inputs.

Emitter Followers

Emitter-follower inputs are used throughout the interface. These are the same as the followers shown in Fig. F-10.

Power Supplies

The 5-V logic is supplied from a Hewlett-Packard model 62005A-011 power supply. The ±15 V are obtained from an Analog Devices Corp. model 903.

Appendix G

Interpreting Specifications Found on Digital Integrated-Circuit Data Sheets

In order to perform the experiments in this manual, it will be necessary to read and interpret the data sheets of various integrated circuits. This appendix provides sufficient information for students to wire and use most common integrated-circuit packages and all the packages used in this book. The appendix also discusses several specifications which are important from a design standpoint but need not be understood before performing the experiments. (Indeed, several of the experiments were designed to teach these concepts. For instance, a timing diagram will be more meaningful *after* the student has completed Experiments 2 and 3.) Thus, this appendix should be reread after completing Experiments 1 through 5.

There are four types of specifications found in the IC data sheets. These are *functional*, *mechanical*, *environmental*, and *electrical*. The functional specifications are related to what the device *does* logically; the mechanical specifications describe the appearance and dimensions of the device and any special mounting requirements; the environmental specifications define the temperature range over which the device can operate satisfactorily; and the electrical specifications describe the power requirements, as well as the electrical features of device operation.

Figure G-1 shows the data sheet for a quadruple 2-input positive NAND gate. The important piece of data for the beginning student is the pin schematic shown in the upper right-hand box. Several different configurations are shown because the package is made in several different forms. Most commonly, we will use epoxy or ceramic-encapsulated dual-in-line packages (DIP) of the A type (Texas Instruments N type). In order to orient the package for correct numbering, each package will have an indentation stamped either above the "one" position or between the first and last pins, 1 and

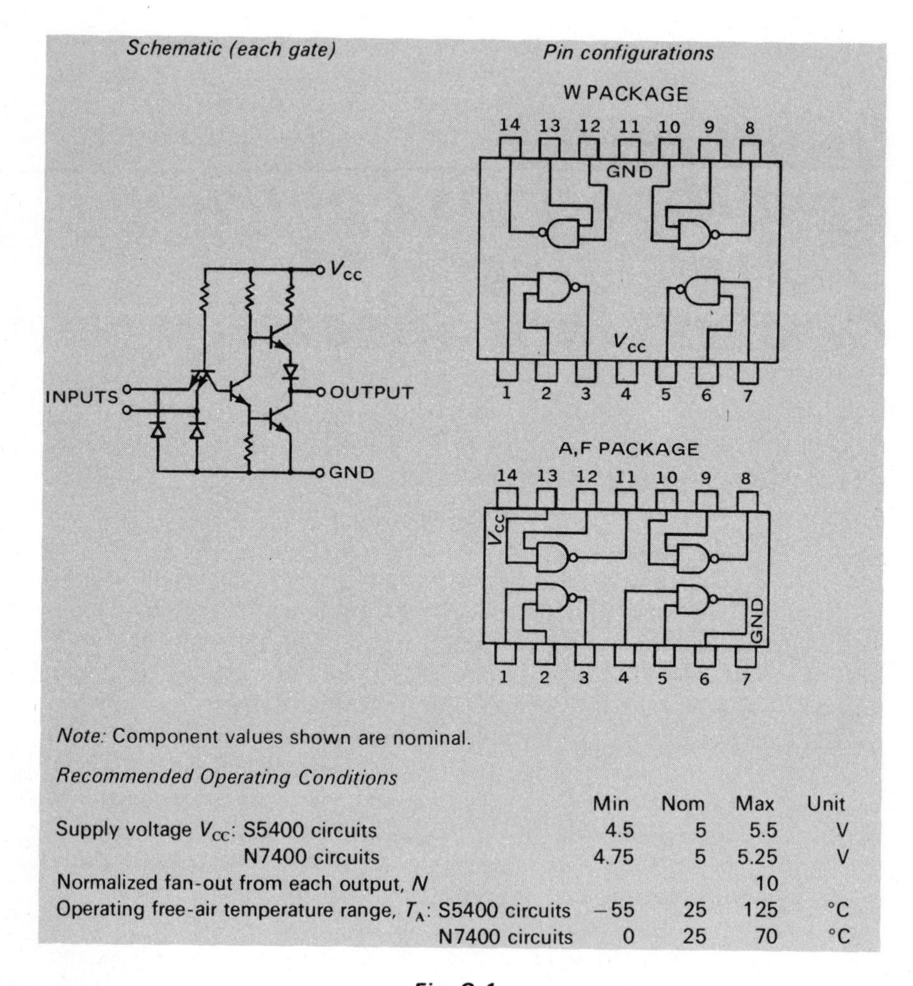

Note: Component values shown are nominal.

Recommended Operating Conditions

	Min	Nom	Max	Unit
Supply voltage V_{CC}: S5400 circuits	4.5	5	5.5	V
N7400 circuits	4.75	5	5.25	V
Normalized fan-out from each output, N			10	
Operating free-air temperature range, T_A: S5400 circuits	−55	25	125	°C
N7400 circuits	0	25	70	°C

Fig. G-1

Typical data sheet for an integrated circuit (courtesy of Signetics Incorporated, Sunnyvale, California).

Electrical characteristics (over recommended operating free-air temperature range unless otherwise noted)

	Parameter	Test Conditions*		Min	Typ**	Max	Unit
$V_{in(1)}$	Logical 1 input voltage required at both input terminals to ensure logical 0 level at output	V_{CC} = min		2			V
$V_{in(0)}$	Logical 0 input voltage required at either input terminal to ensure logical 1 level at output	V_{CC} = min				0.8	V
$V_{out(1)}$	Logical 1 output voltage	V_{CC} = min, $I_{load} = -400\ \mu A$	$V_{in} = 0.8\ V$,	2.4	3.3		V
$V_{out(0)}$	Logical 0 output voltage	V_{CC} = min, $I_{sink} = 16\ mA$	$V_{in} = 2\ V$,		0.22	0.4	V
$I_{in(0)}$	Logical 0 level input current (each input)	V_{CC} = max	$V_{in} = 0.4\ V$			−1.6	mA
$I_{in(1)}$	Logical 1 level input current (each input)	V_{CC} = max	$V_{in} = 2.4\ V$			40	μA
		V_{CC} = max	$V_{in} = 5.5\ V$			1	mA
I_{OS}	Short circuit output current†	V_{CC} = max	S5400	−20		−55	mA
			N7400	−18		−55	
$I_{CC(0)}$	Logical 0 level supply current	V_{CC} = max	$V_{in} = 5\ V$		12	22	mA
$I_{CC(1)}$	Logical 1 level supply current	V_{CC} = max	$V_{in} = 0$		4	8	mA

Switching characteristics, $V_{CC} = 5\ V$, $T_A = 25°C$, $N = 10$

	Parameter	Test Conditions		Min	Typ	Max	Unit
$t_{pd(0)}$	Propagation delay time to logical 0 level	$C_L = 15\ pF$	$R_L = 400\ \Omega$		7	15	ns
$t_{pd(1)}$	Propagation delay time to logical 1 level	$C_L = 15\ pF$	$R_L = 400\ \Omega$		11	22	ns

* For conditions shown as min or max, use the appropriate value specified under recommended operating conditions for the applicable device type.

** All typical values are at $V_{CC} = 5\ V$, $T_A = 25°C$.

† Not more than one output should be shorted at a time.

14 in this case. (The view is *from the top*; that is, the pins are pointing *into the plane of the paper*.) After the chip has been oriented with this indentation on the *left-hand side* (and the pins pointing underneath), the numbering will proceed as shown in the figure; that is, for an N-pin chip, pin 1 is always on the lower left, pin $N/2$ on the lower right, pin $(N/2) + 1$ on the upper right, and pin N on the upper left. Note that *both 14- and 16-pin chips* are used in this book, so that sometimes the lower right-hand pin will be 7, and other times it will be 8. It is important to check pin designations carefully to avoid mistakes which may damage or destroy the integrated circuits. (If a circular package is used, pin No. 1 will be marked, and the numbering then proceeds *counterclockwise* around the chip, again looking from the top.)

For the A package IC shown, pins 1 to 6 and 8 to 13 represent a mixture of inputs and outputs for the four gates contained in the chip. The gate notation is the standard notation, which is introduced in Sec. I and Experiment 1. Thus, pin numbers 1 and 2 are inputs to a particular NAND gate, whose output appears at pin number 3.

Pin number 14 is labeled V_{CC}; V_{CC} is the voltage which must be applied to the chip in order to supply power to the gates. Pin 7 is labeled GND (ground), and provides a return path to the power supply. (For the simple applications of Experiments 1 to 3, ground will be the negative terminal of the 5-V logic power supply or battery. Later experiments requiring positive and negative power supplies are somewhat more complex and should be discussed by the instructor.) Ground is also called "common" because most devices share a common return path to the power supply. One frequent student error is to forget to connect either V_{CC} or GND. The gates will not function in this case.

The next most important specification is the part labeled "Recommended Operating Conditions." Thus, the recommended supply voltage for the 7400 is 5.00 ± 0.25 V. Voltages below 4.75 V may yield spurious results, while those above 5.25 V may destroy the IC. The more expensive 5400 has a slightly greater tolerance and is generally used in military or aeronautical applications where component failure cannot be tolerated.

The circuit schematic at the upper left is quite useful to the designer, as are the *electrical and switching characteristics*. Some electrical specifications are unique for the particular type of device. For example, a one-shot (Experiment 3) will have a specification of the minimum delay time, but this information would be irrelevant for a JK flip-flop (Experiment 2). On the other hand, there are some electrical specifications which are included for all digital IC's, regardless of function. These include: noise margin, fan-out, operating speed, and power dissipation.

NOISE MARGIN

This characteristic is the key to the fidelity of digital logic, yet it is not always explicitly stated in the data sheets. This information can always be obtained, however, from other data-sheet information. Always specified are $V_{in}(1)$, $V_{in}(0)$, $V_{out}(1)$, and $V_{out}(0)$, where in = input; out = output; 1 = high, or on; and 0 = low, or off. (The notation may vary from manufacturer to manufacturer.) The noise margin is calculated from $V_{out}(1)_{min} - V_{in}(1)_{min}$, or from $V_{in}(0)_{max} - V_{out}(0)_{max}$. For example, in the TTL 7400-series IC's referred to in this lab manual, the guaranteed noise margin is 0.4 V (see Fig. G-1).

FAN-OUT

The fan-out of a given IC is the number of parallel inputs which can be driven by a single output. For example, with the 7400 series, an input in the high state requires a 40 μA current, while an output in the high state (I_{load}) can provide 400 μA maximum current. Thus, the fan-out specification is 400/40 = 10.

OPERATING SPEED

This specification defines the time required for a digital IC device to transmit the result of new input data to the output, that is, the time interval that is required between the time a logic level change is seen at the input and the time the corresponding level change appears at the output. This time interval is usually referred to as the "propagation delay time." On the data sheets the specifications given are: $t_{pd}(1)$ (time required to cause the output to change from a low to high level), and $t_{pd}(0)$ (time required to cause the output to change from a high to low level).

POWER DISSIPATION

This specification really refers to the amount of power "used up" by the IC device, that is, of all the power applied to the device, how much is *consumed* by conversion to heat. This value is related to the total number of gates in the device. For example, a value of about 10 mW per gate is the typical power dissipation for a TTL 7400-series IC device.

Appendix H

Operational Amplifiers

Operational amplifiers are used in a variety of applications. In these experiments, the most common applications are dc voltage amplification, current-to-voltage conversion, and impedance matching or circuit-component isolation. These functions, as well as several additional uses, are briefly discussed in this appendix. Because high-gain, low-drift, dc operational amplifiers are available as low cost, IC packages, our emphasis will be on *applications* of these amplifiers, rather than on the basic principles of their operation.

The basic symbol and general characteristics of an operational amplifier are shown in Fig. H-1. The power connections, $\pm V_{CC}$, are usually omitted and will not appear on any other circuits shown in this appendix. The input labeled "$V_{-\text{in}}$" is called the "inverting input," while that labeled "$V_{+\text{in}}$" is called the "noninverting input." Frequently, the noninverting input is grounded, and only the inverting input is shown; if only one input is shown, it is assumed to be the inverting input.

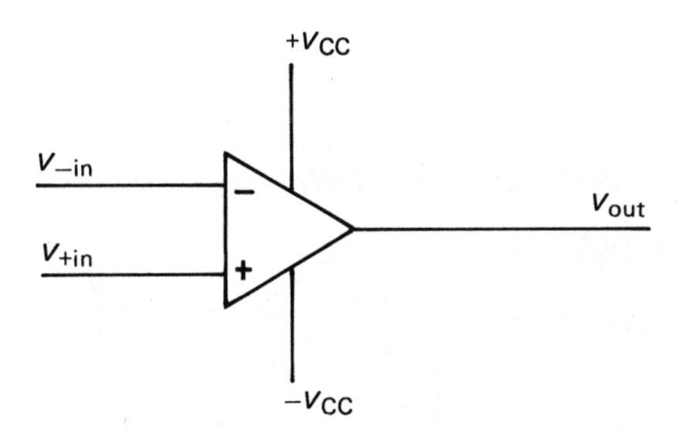

Fig. H-1

Basic symbol and general characteristics for an operational amplifier. $A = \text{gain} = V_{out}/\Delta V_{in} \simeq 10{,}000$ to $100{,}000$.

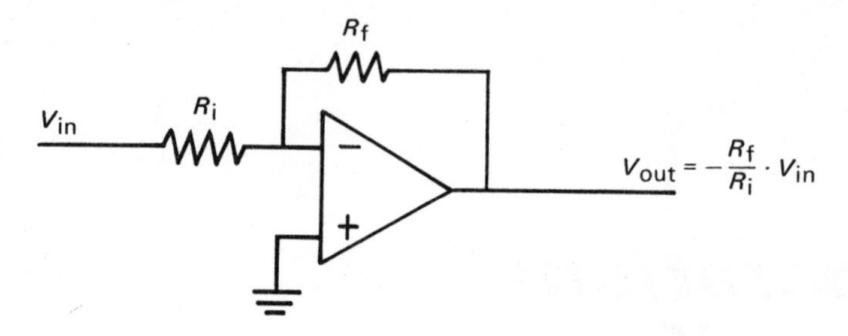

Fig. H-2
A simple voltage amplifier.

VOLTAGE AMPLIFIERS

Figure H-2 shows an operational amplifier wired as a *voltage amplifier.* Note that the sign of the amplified voltage is *opposite* that of the original signal. This will usually be the case when the inverting input is used. It is wise to restrict the gain of any one stage (i.e., the ratio R_f/R_i) to less than 1000 in order to avoid the possibility of oscillation.

With slight modifications, this circuit can be used to sum several signals (Fig. H-3) or as a ranging amplifier (Fig. H-4). It is also possible to filter out ac components by placing a capacitor in parallel with R_f (Fig. H-5). (A wide variety of filter circuits exist; the reader should consult Bibliography A for references containing further discussions of these.) The time constant of the filter is given by the product of R_f and C.

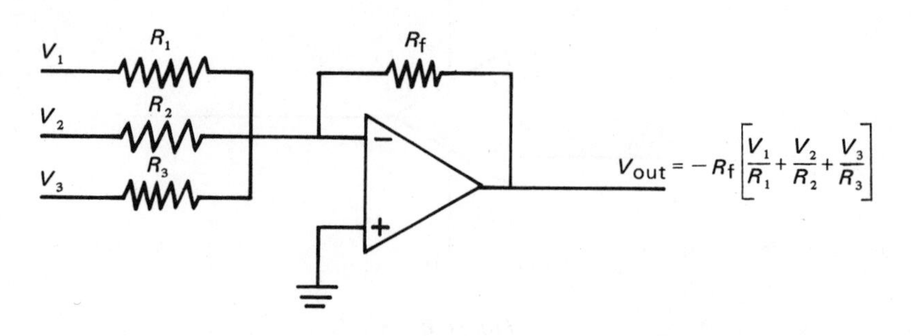

Fig. H-3
A voltage summer.

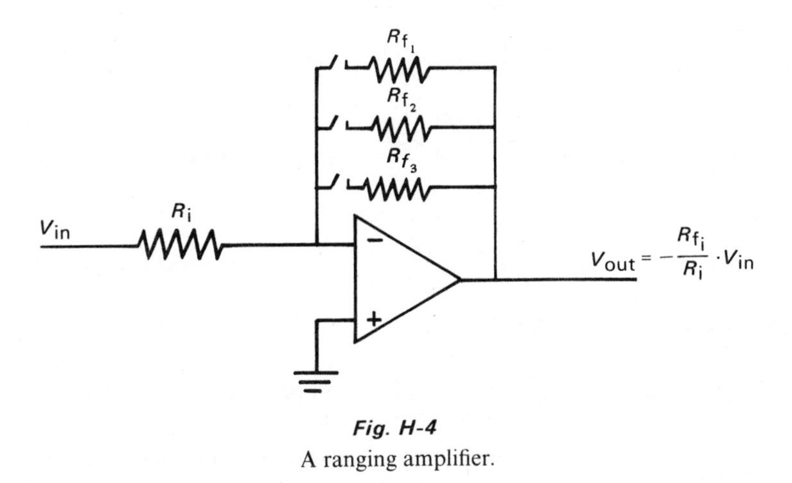

Fig. H-4
A ranging amplifier.

CURRENT-TO-VOLTAGE CONVERTERS

If the input resistor R_i is omitted, then the amplifier circuits H-1 to H-5 act as *current-to-voltage converters.* The output voltage V_{out} will be the (negative) product of the feedback resistor R_f and the total current which passes through that resistor (which is dependent on the output impedance of the device connected to the amplifier inputs). Certain DAC's operate in the current mode and require current-to-voltage conversion in these experiments.

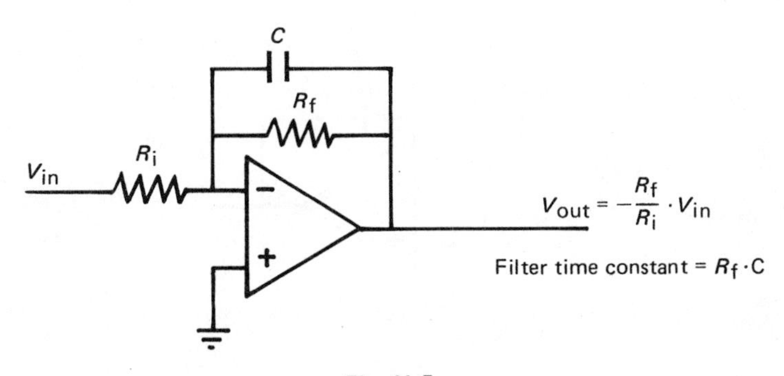

Fig. H-5
A voltage amplifier with filtering.

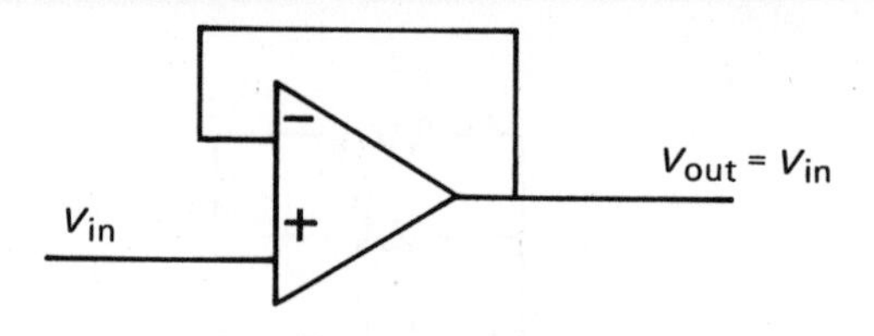

Fig. H-6
A voltage follower.

VOLTAGE FOLLOWERS

Frequently, it is necessary to *isolate* a component in a circuit, so that its operation does not perturb other sections of the circuit. For example, if an analog-to-digital converter is attached to the terminals of a 1-mV recorder, it is quite possible that operation of the ADC may cause the recorder pen to move (hence, generating "spurious" peaks). The device most commonly used to provide this isolation is a *voltage follower*, shown in Fig. H-6. Notice that the output voltage has the *same* sign as the input signal, unlike the voltage amplifier and current-to-voltage converter. As the follower is shown here, it has *unity gain*, and V_{out} should exactly equal V_{in}. It is also possible to design followers with greater than unity gain and which also provide filtering.

INTEGRATORS AND DIFFERENTIATORS

Two other common applications of operational amplifiers are as *integrators* and *differentiators*. These are shown in Figs. H-7 and H-8, respectively.

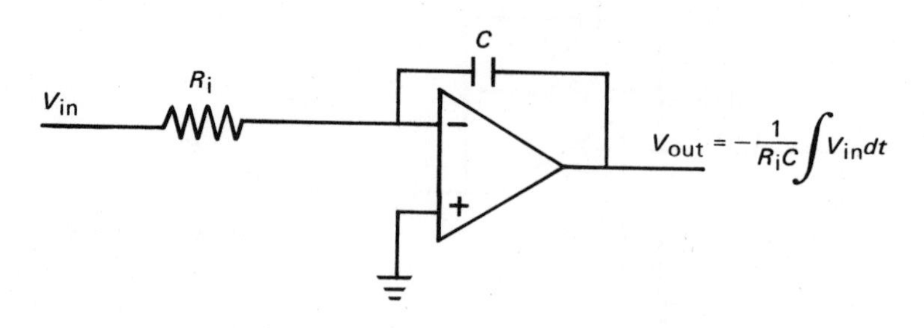

Fig. H-7
An operational amplifier integrator.

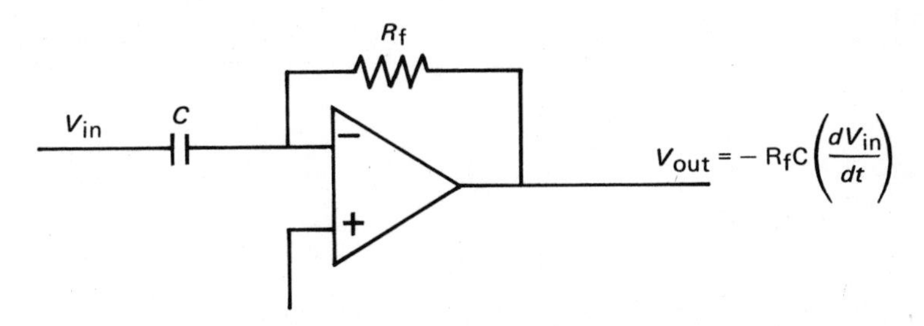

Fig. H-8
An operational amplifier differentiator.

An integrator is used in Experiment 5, with the continuous-ramp comparison ADC; the differentiator is included for completeness, but is not used in these experiments.

Appendix I

Bibliography

A. ELECTRONICS AND DIGITAL LOGIC

A-1. Leonard W. Bell and John W. Sheppard, *Digital Concepts* (Tektronix, Inc., Beaverton, Oregon, 1968).

A-2. R. Ralph Benedict, *Electronics for Scientists and Engineers* (Prentice-Hall, Englewood Cliffs, New Jersey, 1967).

A-3. Edward L. Braun, *Digital Computer Design* (Academic Press, New York, 1963).

A-4. Yaohan Chu, *Digital Computer Design Fundamentals* (McGraw-Hill, New York, 1962).

A-5. Louis A. Delhom. *Design and Application of Transistor Switching Circuits* (McGraw-Hill, New York, 1968).

A-6. A. J. Diefenderfer, *Principles of Electronic Instrumentation* (Saunders, Philadelphia, 1968).

A-7. Angelo C. Gillie, *Binary Arithmetic and Boolean Algebra* (McGraw-Hill, New York, 1962).

A-8. R. G. Hibbard, *Solid-State Electronics* (McGraw-Hill, New York, 1968).

A-9. G. E. Hoernes and M. F. Heilweil, *Introduction to Boolean Algebra and Logic Design* (McGraw-Hill, New York, 1964).

A-10. John L. Hughes, *Computer Lab Workbook* (Digital Equipment Corp., Maynard, Massachusetts, 1969).

A-11. Donald M. Hunter, *Introduction to Electronics* (Holt, New York, 1964).

A-12. Gerald A. Maley and John Earle, *The Logic Design of Transistor Digital Computers* (Prentice-Hall, Englewood Cliffs, New Jersey, 1963).

A-13. H. V. Malmstadt and C. G. Enke, *Digital Electronics for Scientists* (Benjamin, New York, 1969).

A-14. H. V. Malmstadt, C. G. Enke, and E. C. Toren, Jr., *Electronics for Scientists* (Benjamin, New York, 1963).

A-15. A. P. Malvino, *Transistor Circuit Approximations* (McGraw-Hill, New York, 1968).

A-16. C. S. Meyer, D. K. Lynn, and D. J. Hamilton, *Analysis and Design of Integrated Circuits* (McGraw-Hill, New York, 1968).

A-17. J. Miller (ed.), *Transistor Circuit Design* (McGraw Hill, New York, 1963).

A-18. R. L. Morris and John R. Miller *Designing with TTL Integrated Circuits* (McGraw-Hill, New York, 1971).

A-19. Lois Nashelsky, *Digital Computer Theory* (Wiley, New York, 1966).

A-20. Sam P. Perone and David O. Jones, *Digital Computers in Scientific Instrumentation* (McGraw-Hill, New York, 1973).

A-21. Texas Instruments Learning Center, *Understanding Solid-State Electronics* (Texas Instruments, Dallas, 1972).

A-22. Basil H. Vassos and Galen W. Ewing, *Analog and Digital Electronics for Scientists* (Wiley, New York, 1972).

A-23. D. G. Larsen and P. R. Rony, *Bugbook I* (EL Instruments, Inc, Derby, Conn., 1974).

A-24. D. G. Larsen and P. R. Rony, *Bugbook II* (EL Instruments, Inc, Derby, Conn., 1974).

B. BASIC AND REAL-TIME BASIC

B-1. *Programming Languages* (Digital Equipment Corp, Maynard, Massachusetts, 1972).

B-2. Fred Gruenberger, *Computing with the* BASIC *Language* (Canfield Press, San Francisco, 1972).

B-3. *A Pocket Guide to Hewlett-Packard Computers* (Hewlett-Packard, Palo Alto, California, 1970).

B-4. John G. Kemeny and T. E. Kurz, BASIC *Programming* (Wiley, New York, 1967).

B-5. C. E. Klopfenstein and C. L. Wilkins (eds.), *Computers in Chemical and Biochemical Research* (Academic Press, New York, 1972).

B-6. Sam P. Perone and J. F. Eagleston, Introduction of digital computers into the undergraduate laboratory, *J. Chem. Educ.* **48**, 317 (1971).

B-7. *Varian Software Handbook* (Varian Data Machines, Irvine, California, 1971).

B-8. Charles L. Wilkins and Charles E. Klopfenstein, Laboratory computing with Real-Time BASIC, *Chem. Tech.* **2**, 560, 681 (1972).

B-9. Charles L. Wilkins, Charles E. Klopfenstein, Thomas L. Isenhour, Peter C. Jurs, with James S. Evans, and Robert C. Williams, *Introduction to Computer Programming for Chemists*, BASIC *Version* (Allyn and Bacon, Boston, 1975).

C. INTERFACING

C-1. Raymond E. Dessy and Jonathan A. Titus, Computer interfacing, *Anal. Chem.* **45**, 124A (1973).

C-2. *Small Computer Handbook* (Digital Equipment Corp., Maynard, Massachusetts, 1972).

C-3. Brian K. Hahn and C. G. Enke, Real-Time clocks for laboratory-oriented computers, *Anal. Chem.* **45**, 651A (1973).

C-4. *A Pocket Guide to Interfacing HP Computers* (Hewlett-Packard, Palo Alto, California, 1970).

C-5. David F. Hoeschele, Jr., *Analog-to-Digital/Digital-to-Analog Conversion Techniques* (Wiley, New York, 1968).

C-6. G. A. Korn and T. M. Korn, *Electronic Analog and Hybrid Computers* (McGraw-Hill, New York, 1964).

C-7. R. A. Leskovec and P. R. Bevington, A solid state light pen and computer interface for the Tektronix 611 storage oscilloscope, *Rev. Sci. Instr.* **42**, 602 (1971).

C-8. Nick Stadtfeld and Bob LeBrun, *Information Display Concepts* (Tektronix, Inc., Beaverton, Oregon, 1968).

C-9. *Varian 620/L Computer Handbook* (Varian Data Machines, Irvine, California, 1971).

C-10. Rodney C. Williams and T. J. Turner, A simple computer-spectrophotometer interface, *Rev. Sci. Instr.* **43**, 1207 (1972).

D. DIGITAL SMOOTHING AND SIGNAL CONDITIONING

D-1. P. R. Bevington, *Data Reduction and Error Analysis for the Physical Sciences* (McGraw-Hill, New York, 1969).
D-2. W. E. Groves, *Brief Numerical Methods* (Prentice-Hall, Englewood Cliffs, New Jersey, 1966).
D-3. Gary Horlick and Howard V. Malmstadt, Basic and practical considerations for sampling and digitizing interferograms generated by a Fourier transform spectrometer, *Anal. Chem.* **42**, 1361 (1970).
D-4. P. C. Kelly and Gary Horlick, Practical considerations for digitizing analog signals, *Anal. Chem.* **45**, 518 (1973).
D-5. Shan S. Kuo, *Computer Applications of Numerical Methods* (Addison-Wesley, Reading, Massachusetts, 1972).
D-6. B. P. Lathi, *Communication Systems* (Wiley, New York, 1968).
D-7. J. Mandel, *The Statistical Analysis of Experimental Data* (Wiley, New York, 1964).
D-8. A. Ralston, *A First Course in Numerical Analysis* (McGraw-Hill, New York, 1964).
D-9. Abraham Savitzky and Marcel J. E. Golay, Smoothing and differentiation of data by simplified least squares procedures, *Anal. Chem.* **36**, 1627 (1964). (See also the corrections noted in D-10.)
D-10. Jean Steinier, Yves Termonia, and Jules Deltour, Comments on smoothing and differentiation of data by simplified least square procedure, *Anal. Chem.* **44**, 1906 (1972).

E. EXPERIMENTS

Experiment 15

E-1. S. P. Perone and J. F. Eagleston, *J. Chem. Educ.* **48**, 438 (1971).
E-2. O. E. Schupp, III, in *Technique of Organic Chemistry*, Vol. 13, E. S. Perry and A. Weissberger, eds. (Interscience, New York, 1968).
E-3. R. D. McCullough, *J. Gas Chrom.* **5**, 635 (1967).
E-4. F. Baumann, A. C. Brown, and M. B. Mitchell, *J. Chrom. Sci.* **8**, 20 (1970).
E-5. H. A. Hancock, Jr., L. A. Dahm, and J. F. Muldoon, *J. Chrom. Sci.* **8**, 57 (1970).

Experiment 16

E-6. D. O. Jones, M. D. Scamuffa, L. S. Portnoff, and S. P. Perone, *J. Chem. Educ.* **49**, 717 (1972).
E-7. H. O. Malmstadt and H. L. Pardue, *Anal. Chem.* **33**, 1040 (1961).
E-8. H. L. Pardue, *Anal. Chem.* **35**, 1241 (1963).
E-9. H. L. Pardue, M. F. Burke, and D. O. Jones, *J. Chem. Educ.* **44**, 684 (1967).
E-10. E. C. Toren, *J. Chem. Educ.* **44**, 172 (1967).
E-11. H. L. Pardue, *Record Chem. Progr.* **27**, 151 (1966).
E-12. G. G. Guilbault, *Anal. Chem.* **38**, 527R (1966).

Experiment 17

E-13. F. G. Pater and S. P. Perone, *J. Chem. Educ.* **50**, 428 (1973).
E-14. P. Job, *Ann. Chim.* (*Paris*) [10]**9**, 113 (1928).
E-15. W. C. Vosburgh and G. R. Cooper, *J. Am. Chem. Soc.* **63**, 437 (1941).

Index